Emissions and Air Quality

Other SAE books on this topic:

Automobiles and Pollution
by Paul Degobert
(Order No. R-150)

Emissions from Two-Stroke Engines
by Marco Nuti
(Order No. R-223)

Reduced Emissions and Fuel Consumption in Automobile Engines
by F. Schäfer and R. van Basshuysen
(Order No. R-157)

For more information or to order this book, contact SAE at
400 Commonwealth Drive, Warrendale, PA 15096-0001;
phone (724) 776-4970; fax (724) 776-0790; e-mail: publications@sae.org.

Emissions and Air Quality

Hans Peter Lenz
Christian Cozzarini

Society of Automotive Engineers, Inc.
Warrendale, Pa.

Library of Congress Cataloging-in-Publication Data

Lenz, Hans Peter,
Emissions and air quality / Hans Peter Lenz, Christian Cozzarini
p. cm.
Includes bibliographical references and index.
ISBN 0-7680-0248-6
1. Motor vehicles--Motors--Exhaust gas--Environmental aspects. 2. Air quality. I. Cozzarini, Christian. II. Title.
TD886.5.L46 1999
363.738'7--dc21 99-14564
CIP

400 Commonwealth Drive
Warrendale, PA 15096-0001 U.S.A.
Phone: (724) 776-4841
Fax: (724) 776-5760
E-mail: publications@sae.org
http://www.sae.org

ISBN 0-7680-0248-6

SAE Order No. R-237

Preface

The work presented here summarizes the results of a scientific study performed by Prof. Dr. H.P. Lenz and Dipl.-Ing. Christian Cozzarini from the Institut für Verbrennungskraftmaschinen und Kraftfahrzeugbau at the Technische Universität Wien. The comprehensive text of all relevant work is contained in the studies "Zur Bedeutung der Stickoxide" ("About the Importance of Nitrogen Oxides") and "Emission, Immission, und Wirkung von Abgasbestandteilen" ("Emissions, Air Quality, and Effects of Exhaust Gas Constituents") published in 1993, 1994, 1995, 1996, and 1997 [40, 47, 60, 143–145], which summarize the present state of scientific knowledge based on more than 1,500 literature sources.

Many thanks are extended to the following companies for providing the financial support that made this work possible:

Adam Opel AG, Rüsselsheim
Daimler-Benz AG, Stuttgart
MAN AG, Nürnberg
Volkswagen AG, Wolfsburg
BMW AG, München
Ford-Werke AG, Köln
Dr.-Ing. h.c. F. Porsche AG, Stuttgart

Special thanks go to Dr.-Ing. Wolfgang Berg (Daimler-Benz), Dipl.-Ing. Marc Boderke (Daimler-Benz), Prof.-Dr. Walter Brandstetter (Ford), Dr.-Ing. Wolfgang Held (MAN), Dr.-Ing. Peter Kohoutek (VW), Dr.-Ing. Norbert Metz (BMW), Dipl.-Ing. Georg Schaffner (Opel), and Dr.-Ing. Dieter Schürmann (VW) as members of the working group that monitored the study under the guidance of Prof. Dr. Dusan Gruden (Porsche). We are grateful for their personal engagement and important assistance.

Hans Peter Lenz
Christian Cozzarini

Vienna, August 1998

Contents

Abstract xi

Chapter 1 Introduction—Objective of the Study 1

Chapter 2 The Basics 3
- 2.1 Effects of Air Polluting Substances on the Atmosphere 3
- 2.2 Evaluation Methodology—Critical Consideration of Results 5
- 2.3 Difference Between Substances with Local and Global Effects on the Atmosphere 5

Chapter 3 Emissions and Atmospheric Concentrations of Substances with Global Effects 7
- 3.1 The Greenhouse Effect 7
- 3.2 Stratospheric Ozone Depletion 12
  - 3.2.1 Water Vapor 13
  - 3.2.2 Carbon Dioxide 14
  - 3.2.3 Halogenated Hydrocarbons 19
  - 3.2.4 Methane 23
  - 3.2.5 Nitrous Oxide 25

Chapter 4 Exhaust Emission Components with Local and Regional Effects 31
- 4.1 Nitrogen Oxides 32
  - 4.1.1 Nitrogen Oxide Emissions in Europe 33
  - 4.1.2 Nitrogen Oxide Emissions in the Federal Republic of Germany 36
  - 4.1.3 Judgment Criteria for the Impact of Nitrogen Oxides on Air Quality 37
  - 4.1.4 Concentrations of Nitrogen Oxides in the Atmosphere 39
- 4.2 Non-Methane Hydrocarbons 40
  - 4.2.1 Non-Methane Hydrocarbon Emissions in Europe 41
  - 4.2.2 Non-Methane Hydrocarbon Emissions in the Federal Republic of Germany 42

4.2.3 Concentrations of Non-Methane Hydrocarbons in the Atmosphere ... 44

4.3 Sulfur Dioxide ... 44

4.3.1 Sulfur Dioxide Emissions in Europe ... 45

4.3.2 Sulfur Dioxide Emissions in the Federal Republic of Germany ... 46

4.3.3 Judgment Criteria for the Impact of Sulfur Dioxide on Air Quality ... 48

4.3.4 Concentrations of Sulfur Dioxide in the Atmosphere ... 50

4.4 Dust and Particulate Matter ... 52

4.4.1 Airborne Dust Emissions in Europe ... 53

4.4.2 Airborne Dust Emissions in the Federal Republic of Germany ... 54

4.4.3 Judgment Criteria for the Impact of Airborne Dust on Air Quality ... 55

4.4.4 Concentrations of Airborne Dust in the Atmosphere ... 58

4.5 Carbon Monoxide ... 60

4.5.1 Carbon Monoxide Emissions in Europe ... 60

4.5.2 Carbon Monoxide Emissions in the Federal Republic of Germany ... 61

4.5.3 Judgment Criteria for the Impact of Carbon Monoxide on Air Quality ... 63

4.5.4 Concentrations of Carbon Monoxide in the Atmosphere ... 63

4.6 Ozone ... 64

4.6.1 Judgment Criteria for the Impact of Ozone on Air Quality ... 65

4.6.2 Concentrations of Ozone in the Atmosphere ... 66

4.7 Benzene ... 68

4.7.1 Benzene Emissions in the Federal Republic of Germany ... 69

4.7.2 Judgment Criteria for the Impact of Benzene on Air Quality ... 69

4.7.3 Concentrations of Benzene in the Atmosphere ... 71

4.8 Polycyclic Aromatic Hydrocarbons ... 72

4.8.1 Concentrations of Polycyclic Aromatic Hydrocarbons in the Atmosphere ... 73

Chapter 5 Possible Ways to Influence Emissions from Road Traffic—Emission Scenarios ... 75
5.1 Calculation Model for Determining Emissions from Passenger Car and Heavy-Duty Vehicle Traffic in the Federal Republic of Germany ... 76
5.2 Legislative Measures ... 83
5.2.1 Level of Emission Standards ... 83
5.2.2 In-Use Vehicle Emissions ... 87
5.2.3 Fuel Composition ... 89
5.2.4 Testing of In-Use Vehicles ... 95
5.3 Vehicle User ... 96
5.4 Traffic Management ... 98

Chapter 6 Literature ... 101

Index ... 115

About the Authors ... 125

Abstract

Based on the 1992 VDI Progress Report "Emissions, Air Quality, and Effects of Exhaust Gas Constituents" [127], this study continues to evaluate the current worldwide state of knowledge about the interrelationship between emissions and air quality. By reviewing more than 1,500 literature sources through 1998, the study describes the contribution of passenger car and commercial vehicle traffic to local and global emission situations, together with the consequences for the environment.

Because of their life span, trace gases emitted into the atmosphere by natural sources and by manmade exhaust gases are distinguished as components with global or regional/local effects. Trace gases reacting on a global scale—such as water vapor, carbon dioxide, methane, and nitrous oxides—are considered responsible for the natural "greenhouse effect" on the earth. Worldwide anthropogenic emissions—which are emissions from burning fossil fuels in power stations, industry, and home heating, as well as emissions from agriculture and stock farming—contribute 0.5 to 1.5% to this effect.

From these components, water vapor followed by carbon dioxide contribute most to the global greenhouse effect. Carbon dioxide (CO_2) emissions from natural sources represent 96.5% of total worldwide CO_2 emissions, thus exceeding anthropogenic carbon dioxide emissions by orders of magnitude. Worldwide passenger car and commercial vehicle traffic contribute 0.5% to total CO_2 emissions. Although relatively small, this share of manmade CO_2 emissions is nevertheless considered as additional emissions that may adversely affect the natural CO_2 equilibrium and thus global climate.

In addition, methane (CH_4) and nitrous oxides (N_2O) are important greenhouse gases. Anthropogenic methane emissions represent two-thirds of the total worldwide emissions, and manmade nitrous oxides represent one-third of total worldwide emissions. The highest methane emissions stem as much from natural sources such as moors and wetlands as from anthropogenic sources such as wet rice agriculture and stock farming. Methane emissions from road traffic are so marginal that their contribution to the greenhouse effect can be disregarded.

The greatest amounts of nitrous oxides are emitted from forests, grasslands, and oceans. Further emissions are caused by nitrogen-containing fertilizers used in farming. Passenger car and heavy-duty vehicle traffic each contribute 1.6% to the total anthropogenic emissions of nitrous oxide. Therefore, we can conclude that the contribution of road traffic to emissions with global air quality effects is very small.

Regarding emissions that may cause regional and local effects on air quality close to the area where they are emitted, the share contributed by the individual source is not the only consideration. It is also important to compare the existing absolute level against legally prescribed air quality standards and to evaluate the trend of these emissions for future years. Emissions with regional and local effects are carbon monoxide, non-methane hydrocarbons, nitrogen oxides, sulfur dioxide, and suspended particulate matter. Concerning these components, the study provides an overview of the relevant situation within Europe, particularly for the Federal Republic of Germany.

The evaluation shows that, depending on the exhaust substance considered, road traffic emissions already meet existing air quality demands in many cases. For those that do not, they soon will comply with such standards when vehicles with emission control technology adhering to the latest legislative requirements become widely used and replace "old" vehicles without emission control equipment. In addition, the calculations performed demonstrate that a distinct downward trend can be expected for all these substances in the years to come. These conclusions are important *vis-à-vis* the fact that road traffic, when considered on a relative basis, contributes 30 to 50% of the total emissions of nitrogen oxides, carbon monoxide, and benzene in the European Union and Germany.

With regard to non-methane hydrocarbon emissions, as much as one-third of all non-methane hydrocarbons stem from natural sources, namely, forests. The primary emitters of sulfur dioxide emissions are power stations and heating stations. These contribute 50 to 60%, whereas road traffic contributes only about 1%. Concerning suspended particulate matter, more than 80% of total emissions come from industrial activities; road traffic contributes less than 10%.

Finally, the study describes some possibilities for reducing emissions from road traffic through technical and legislative measures. It also quantifies their effect and presents a summary of the findings. Regarding the impact of emission control legislation, the most important result of the calculations is that the step from EU3 to EU4 standards results in only a marginal effect on road traffic emission reduction or, in other words, on air quality. Substantial emission reductions also can be achieved by traffic management, which allows for smooth riding and avoids energy-wasting and emission-causing stop-and-go conditions.

In the context of finding an effective means for in-field emission reductions for the total car population in the Federal Republic of Germany, this study highlights the important role that must be attributed to improved fuel. It shows that particulate matter emissions could be reduced by up to 20% from passenger cars and up to 10% from trucks if modified fuels (i.e., with 50 ppm sulfur content) were used. The use of low aromatic and low benzene fuels could reduce by 50% the benzene emissions from all passenger car traffic immediately on availability of such fuels.

The study concludes by reminding readers that car users can contribute to the reduction of emissions from road traffic and to the reduction of fuel consumption by meaningful operation and good maintenance of their vehicles.

CHAPTER 1

Introduction—Objective of the Study

This study scientifically reviews the current state of knowledge about air pollution caused by mobile sources. The objective of this investigation was to question the perception frequently triggered in the public that air pollution is almost exclusively caused by road traffic.

First, the study evaluates global and regional effects of exhaust emissions from road traffic on the atmosphere, and it investigates the impact of globally acting exhaust substances on the depletion of stratospheric ozone, or the greenhouse phenomenon. With emphasis on the European situation, it further establishes an overview of the contribution of different emission sources to the total level of given emission components. The study compares existing air quality levels with legislative standards established for protecting the environment and population, in order to put the findings about absolute values and relative shares into an accurate perspective. The picture about levels and effects of emissions from road traffic is completed by showing not only historical developments but also (and most importantly) future trends for those exhaust gas constituents that remain under consideration for further legislative measures.

By means of a computer program developed at the Technical University of Vienna, the study calculates the effect of different emission control measures. It highlights the fact that emission reduction is not solely a technical challenge for the automobile and oil industry, but that an important share also can be contributed by vehicle users and intelligent traffic management.

CHAPTER 2

The Basics

The atmosphere of the earth is composed of 21% oxygen and 78% nitrogen, as well as water vapor and trace gases. For millions of years, this atmospheric composition has been subjected to large or small variations. This is also true for trace gases such as carbon dioxide (CO_2) water vapor, ozone (O_3), nitrous oxide (N_2O), methane (CH_4), and various others. The composition of the atmosphere is the result of self-stabilizing interactive effects, which lead to a somewhat indifferent equilibrium. Within this environment, natural catastrophes such as volcanic eruptions can be compensated for without causing a collapse of the ecosystem. However, during the last 100 years, a change with regard to trace gases has been monitored, which is evolving at increasing speed. This change is being attributed to anthropogenic activities [128–130]. The phenomenon might be explained by the world population explosion and the resulting increase in energy consumption, as well as by increasing industrialization and agriculture.

2.1 Effects of Air Polluting Substances on the Atmosphere

Components that change the natural composition of the atmosphere are considered air polluting substances. In this context, it is not important whether these substances exist in a gaseous, liquid, or solid state, or whether they are emitted by natural or anthropogenic sources [84].

A complex sequence of physical and chemical processes occurs between the emission of substances and their deposition on various kinds of surfaces. These processes explain why the concentrations of such substances in the air always vary with location and time. The path of a substance through the atmosphere is characterized by the following steps:

emission $\rightarrow$ transportation $\rightarrow$ transformation $\rightarrow$ concentration $\rightarrow$ deposition

Emission sources can be grouped into two different categories:

1. Natural sources
2. Anthropogenic sources

Natural sources of atmospheric trace substances are fauna and flora, volcanoes, the surfaces of the oceans and the continents, and atmospheric events such as lightning. With regard to anthropogenic sources, all human activities such as power generation, industry and business, traffic, home heating, waste incineration, dumping, agriculture, and (strictly speaking) human breathing must be considered.

After leaving their sources, the emissions mix with ambient air. There they are somewhat diluted before being transported by air streams over small or large distances. During this transportation, the substances will be transformed by physical and chemical processes. Physical processes cause an exchange of substances among airborne particles (aerosols), cloud droplets, fog droplets, raindrops, and the surrounding gas phase. Chemical processes change the trace gas mixture through homogeneous and heterogeneous reactions. Homogeneous chemical reactions are those to which only gaseous reaction partners contribute. Heterogeneous reactions occur on the surfaces of aerosol particles, cloud droplets, fog droplets, or raindrops. Reactions also occur inside these cloud droplets, fog droplets, and raindrops.

The chemical transformation of atmospheric trace substances leads to products such as carbon dioxide and water, which generally are harmless to human beings. However, it also creates some secondary air pollutants. The "motor" for many of these processes is radiation from the sun.

Air quality generally is understood as the local concentration of trace substances in the atmosphere. These concentrations are influenced by local emissions via transportation, physical and chemical transformations, and deposition processes. Because these processes are subject to variations over time, air quality likewise is highly variable and frequently follows a pronounced daily cycle and a distinct yearly cycle.

2.2 Evaluation Methodology—Critical Consideration of Results

Data about emissions from natural and anthropogenic sources vary largely in the existing literature. The most probable values for global anthropogenic and natural emissions mentioned in this study represent a judgment of results from various authors. If results were contradictory, only the latest publications were taken into account, as long as they had adequately considered preceding literature and therefore could be regarded as plausible. Literature sources that independently calculated identical emission values support the most probable value.

The great uncertainties, which became obvious with regard to the power of the individual sources, lead to a data presentation in the form of ranges rather than distinct values. These ranges were not simply taken from individual publications that provided data about different sources; rather, they represent the minimum and maximum values found in all the literature reviewed. Implausible data were not included in the evaluation. In most cases, the ranges are not symmetrical to the most probable values that represent the state of knowledge of the mid-1990s.

To reduce the probability of error, data about so-called mixed natural/anthropogenic sources were equally attributed to each of the two categories. For example, this was done for the oxidation of ammonia, which is by itself a natural process, whereas the participating substances are of different origin. Chemical reactions in the air (i.e., the role of the atmosphere as a secondary emission source) were taken into account as much as possible from the data given in the literature. The secondary products were attributed to the natural or anthropogenic source category according to their basic origins.

2.3 Difference Between Substances with Local and Global Effects on the Atmosphere

A general distinction is made between substances having a long or short lifetime in the atmosphere.

Long-Life Components with Global Effects

Components such as carbon dioxide, methane, halogenated hydrocarbons, and nitrous oxide remain in the atmosphere for several years or even several hundred years. These components are well mixed and are distributed homogeneously throughout the atmosphere. Therefore, only a few measuring stations are needed on the earth to determine the average concentration of these gases in the troposphere. These substances influence the entire atmosphere (i.e., the whole climate), and thus they contribute to the greenhouse effect or change the protective ozone layer in the stratosphere.

Short-Life Components with Local Effects

Short-life components such as carbon monoxide, nitric oxide, nitrogen oxide, and various hydrocarbons may have a lifetime ranging from only several hours to a maximum of several months. The atmospheric concentrations differ widely and depend largely on the geographic location of the source from which they are emitted. These substances directly influence air quality in close vicinity to the source. The short-life components generally will be transformed within a short time into water-soluble end products, which then will be washed out by precipitation. However, this definition of short-life components is not always valid. When these substances are transported to high altitudes or over long distances by certain air streams, they may have a regional or even global effect on air quality.

CHAPTER 3

Emissions and Atmospheric Concentrations of Substances with Global Effects

Gases that produce global effects influence the atmosphere by increasing the greenhouse effect or by changing the ozone layer in the stratosphere under certain circumstances. Components such as chlorofluorocarbons reduce the ozone layer and increase the greenhouse effect.

3.1 The Greenhouse Effect

The atmosphere of the earth receives energy from sun radiation. For the earth to remain in an energetic equilibrium, this energy must be returned to space by certain means. This occurs when the incoming short-wave radiation on the surface of the earth is transformed into long-wave heat radiation, which then is transmitted into space.

The greenhouse effect results because the absorption characteristics of certain components of the atmosphere are not identical in different areas of the spectrum. Thus, incoming light is only partially absorbed, whereas outgoing heat radiation will more or less be absorbed by several trace gases. This means that the necessary discharge of energy is being hampered.

The resulting temperature increase on the surface of the earth is called the greenhouse effect and is a prerequisite for life on our planet. Without the natural greenhouse effect, the earth would have an average temperature of −18°C (0°F). Instead, this temperature is 15°C (60°F) because of the trace gases, water vapor, carbon dioxide, methane, nitrous oxide, and ozone contained in the atmosphere in their natural ratios [6].

In the mid-1980s, public attention was drawn increasingly to the trace gases, carbon dioxide (CO_2), methane (CH_4), nitrous oxide (N_2O), and halogenated hydrocarbons because these cause a change in the natural concentration of the atmosphere and thereby contribute to the greenhouse effect. These effects, which are attributed to manmade trace gases, are combined under the term "anthropogenic greenhouse effect."

To limit this anthropogenic greenhouse effect, it was decided during the December 1997 Kyoto Conference to reduce the following six greenhouse gases:

- Carbon dioxide (CO_2)
- Nitrous oxide (N_2O)
- Methane (CH_4)
- Hydrofluorocarbons (HFCs)
- Perfluorocarbons (PFCs)
- Sulfurhexafluoride (SF_6)

These gases were to be reduced on a worldwide basis by at least 5% compared to the 1990 level until the reference time frame of the years 2008 to 2012. For the European Union, the reduction was to be at least 8% [139].

In literature, the anthropogenic share in the total greenhouse effect will be determined by either considering the average temperature increase close to the surface measured over the last 100 years of 0.45°C (33°F) or the calculated increase in terrestrial radiation flow as being of mere anthropogenic origin [17].

Satellite measurements show that the average temperature on earth is decreasing (Figure 1). This kind of measurement does not detect the temperature on the surface but rather from the air column in the troposphere. However, there are indications that these satellite measurements must be corrected, which then would lead also to a slight temperature increase [18].

In relevant literature, the estimates about the degree of this effect vary largely. However, water vapor represents the most important of all natural greenhouse gases (Figure 2).

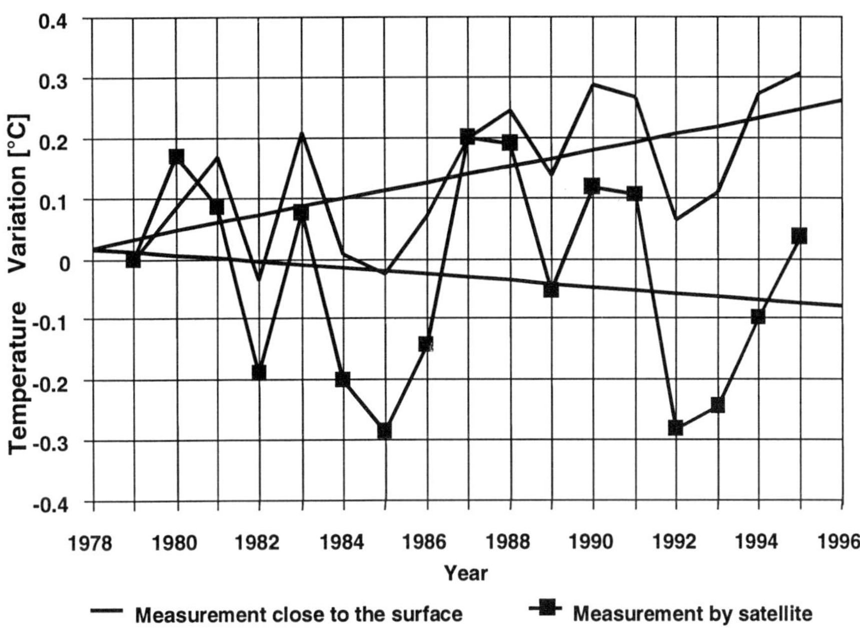

Figure 1. Trend of earth temperature from measurements close to the surface and measurements by satellite [19].

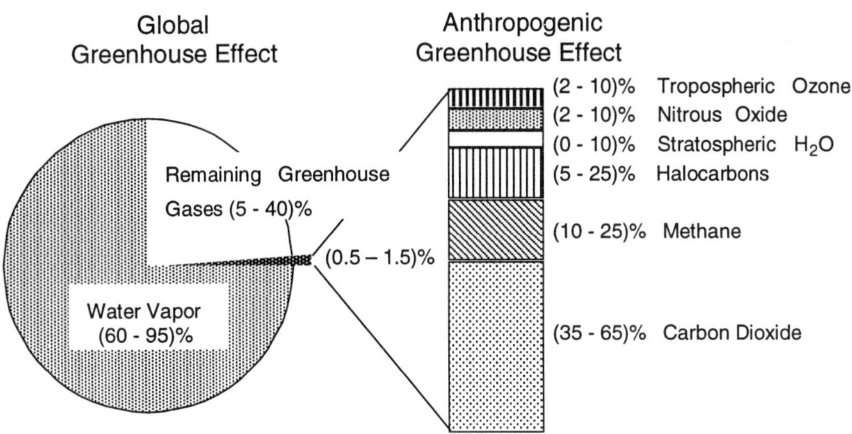

Figure 2. Contribution of anthropogenic greenhouse gases to the natural greenhouse effect [7–16].

The right column of Figure 2 shows that assumptions vary widely about the manmade greenhouse effect which contributes 0.5 to 1.5% to the global greenhouse effect. However, all scientific publications name carbon dioxide as the substance having the largest effect on the anthropogenic greenhouse effect.

To compare emissions of individual greenhouse gases, "CO_2-equivalents" are used (Table 1).

On one hand, these CO_2-equivalents are defined differently by scientists. Likewise, because of the complexity of the processes involved, the CO_2-equivalents are not yet well supported. On the other hand, the orders of magnitude of their values are not questioned. When comparing CO_2-equivalents, the different atmospheric lifetimes of these substances must be taken into account. That is why these CO_2-equivalents depend largely on the time period to which the greenhouse effect is being referred. The CO_2-equivalent referred to as 20 years is an indicator for short-term changes; the CO_2-equivalent referred to as 100 years is an indicator for long-term changes.

Representing total road traffic, Figure 3 depicts the greenhouse potentials of a gasoline engine passenger car without exhaust gas aftertreatment from the year 1975, as well as for another vehicle equipped with a three-way catalyst from the year 1996. The emissions should be understood as representative average values of newly registered gasoline engine passenger cars from the years 1975 and 1996 (see Chapter 5, Section 5.1).

As Figure 3 shows, the greenhouse gases of nitrous oxide emissions and methane emissions play a minor role for modern passenger cars. The "classic" exhaust gas components of carbon monoxide, nitrogen oxides, and unburned non-methane hydrocarbons belong to the so-called indirectly climate-affecting trace gases. These can influence the radiation equilibrium in the troposphere by changing the ozone concentration. The relevance of these components to the climate becomes less important with the successful introduction of the three-way catalyst on passenger cars with gasoline engines and the introduction of low-polluting diesel passenger cars. Carbon dioxide emissions cannot be reduced substantially by means of catalysts. Because these emissions are directly proportional to the amount of fuel used, automobile manufacturers must undertake great efforts to reduce the fuel consumption responsible for

CO_2 emissions by suitable technical measures and further development of engines. Figure 3 demonstrates that these efforts are successful.

TABLE 1. LIFETIME IN THE ATMOSPHERE AND GREENHOUSE POTENTIAL OF TRACE GASES (IN CO_2-EQUIVALENTS) [3, 4, 8, 20, 21]

Component	Concentration in the Atmosphere	Lifetime (Years)	CO_2-Equivalent (Time Horizon 20 Years)	CO_2-Equivalent (Time Horizon 100 years)
H_2O	2 ppm–2%	0.01–2	?	?
CO_2	364 ppm[1996]	50–200	1	1
N_2O	311 ppb[1993]	120–150	270–290	290–320
O_3 (tropospheric)	20 ppb[1993]	0.1–0.2	2,000	2,000
CH_4	1731 ppm[1996]	7–12.3	56–63	21–24
SF_6	3 ppt[1996]	3,200	16,500	24,900
CCl_4	103 ppt[1996]	42–50	1,900	1,300
CH_3CCl_3	98 ppt[1996]	6	350	100
CFCs				
11	272 ppt[1996]	50–60	4,500–5,000	3,500–4,000
12	529 ppt[1995]	102–130	7,100–7,900	7,300–8,500
113	85 ppt[1995]	85–90	4,500–5,000	4,200–5,000
114	20 ppt[1992]	200–300	6,000–6,900	6,900–9,300
115	<10 ppt[1992]	400–1,700	5,500–6,200	6,900
HCFCs				
22	119 ppt[1995]	13.3–15	4,100–4,300	1,500–1,700
123	Presently no data available	1.4–1.6	300–310	85–93
124	Presently no data available	5.9–6.6	1,500	430–480
141b	4 ppt[1995]	8-9.4	1,500–1,600	440–630
142b	7 ppt[1995]	19–19.5	3,700–4,200	1,600–2,000
HFCs				
125	Presently no data available	28–36	4,600–4,700	2,500–2,800
134a	2 ppt[1996]	14–16	3,200–3,400	1,200–1,300
143a	Presently no data available	41-48.3	4,500–5,000	2,900–3,800
152a	Presently no data available	1.5–1.7	460–510	140
Halons				
1211	3 ppt[1996]	20	?	?
1301	2 ppt[1996]	65–110	5,800	5,800

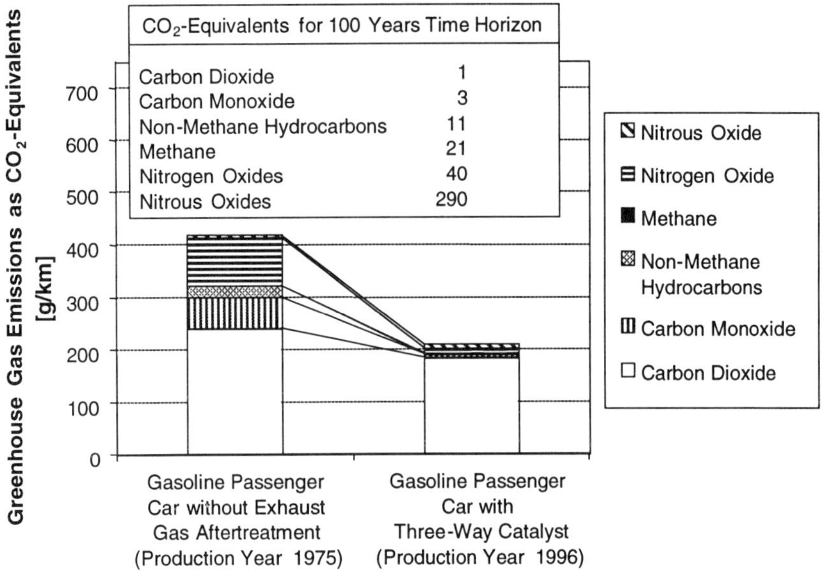

Figure 3. Comparison of greenhouse gas potentials between passenger cars without exhaust gas aftertreatment and passenger cars with three-way catalysts as CO_2-equivalents.

3.2 Stratospheric Ozone Depletion

Approximately 90% of all atmospheric ozone exists in the stratosphere at an altitude of around 20 km (65,000 feet). This ozone layer protects the earth from excessive ultraviolet (UV) radiation through the absorption characteristics of ozone [1].

A dynamic equilibrium among many physical and chemical processes of generation and depletion determines the natural daily and seasonal fluctuations of the ozone concentrations in the atmosphere [132]. However, this equilibrium can be disturbed by anthropogenic emissions. One of the most prevalent theories states that the emission of ozone depleting substances leads to the formation of an "ozone hole" over the Antarctic every year during the months of September and October [2]. The formation of the Antarctic "ozone hole" requires extremely low temperatures, at minimum -70°C (-94°F). Such conditions may occur in the Antarctic during the sunshine-free polar winter season from May to September and the intensive spring sunshine in October. These conditions are a prerequisite for the occurrence of certain reactions on the surface of polar stratospheric clouds. Because of the chemical processes in the atmosphere and high wind speeds (up to 400 km/h [250 mph, or 216 knots]) in the polar air swirl, air masses with high ozone content from tropical regions can hardly intrude in the inner area of the polar air swirl. After the polar swirl collapses during the spring season of the southern hemisphere, the "ozone hole" is filled again within a short time.

Low ozone concentrations above the Antarctic were discovered at the end of the 1950s. The "ozone hole" phenomenon was first detected at the beginning of the 1980s by means of satellite measurements.

The ozone depleting effect of certain trace gases is characterized by the ozone depleting potential (ODP), with CFC 11 being attributed the value 1 within this grouping (Table 2).

3.2.1 Water Vapor

The most important greenhouse gas that is usually omitted from discussion of this phenomenon is water vapor, which contributes approximately 60 to 95% of the total greenhouse effect [7–16]. Finely distributed water vapor in the atmosphere generally occurs as humidity. Water in the atmosphere constitutes a complex dynamic system in which the different states such as vapor, liquid water droplets, and ice crystals interact permanently. Water affects the climate in every physical state in which it exists in the atmosphere. Water vapor that results primarily from the evaporation of ocean surface water passes through a rapid hydrological cycle. Within an average of eight to ten days, it completes this cycle from evaporation to precipitation in the form of rain or snow [3, 7].

Water evaporation draws heat from the air above the ocean surface. This heat will be released when water condenses to droplets in the clouds. Thus, huge amounts of heat are transported through the general circulation within the atmosphere. Depending on the temperature of the surface at a given location, the wind speed, and the humidity, these heat amounts influence the global climate and, in the short term also, the regional weather.

TABLE 2. LIFETIME IN THE ATMOSPHERE AND OZONE DEPLETION POTENTIAL (ODP) OF VARIOUS TRACE GASES [3–5]

Trace Gas	Lifetime (Years)	ODP
Carbon Tetrachloride	50	1.08
1,1,1-Trichloroethane	6	0.12
CFCs		
11	50–60	1
12	102–130	0.9
113	85–90	0.85–1.1
114	200–300	0.8–0.9
115	400–1,700	0.4
HCFCs		
22	13.3–15	0.05
123	1.4–1.6	0.02
124	6.6	<0.2
141b	8–9.4	0.1
142b	19–19.5	0.07
HFCs		
125	28–36	<0.001
134a	14–16	<0.001
143a	41	<0.001
152a	1.7	<0.001
Halons		
1211	20	4.1
1301	65–110	12.5–13

3.2.2 Carbon Dioxide

Carbon dioxide is a colorless and odorless gas. It remains rather inert because CO_2 molecules do not decompose in the atmosphere. Instead, CO_2 will be bound on the surface of plants through absorption of energy (photosynthesis), or it will be stored in the oceans [22]. Today, the atmosphere stores approximately 750 Gt carbon. Oceans bind approximately 40,000 Gt carbon, and the biosphere holds approximately 1,000 Gt carbon [3, 5, 8].

A complete carbon balance exists only in the atmosphere [3, 5, 8]. The influx of carbon dioxide through decay processes, land use, breathing, evaporation from oceans, and burning of fossil fuels, together with less removal of CO_2 from the atmosphere due to photosynthesis and storage in the oceans, results in a differential amount of 3 to 4 Gt C/yr. This amount of carbon must also be stored in the atmosphere, which leads to an increase in the average tropospheric CO_2 concentration of approximately 0.45% per year. Precise direct measurements of atmospheric CO_2 concentrations have existed only since 1959, when measurement stations were established on Mauna Loa (Hawaii) and at the South Pole.

Chemical analyses from air bubbles trapped in glacier ice provided data on the historical development of the pre-industrial average global tropospheric CO_2 concentration (Figure 4, top).

Because of its long life (i.e., its long presence in the atmosphere), carbon dioxide is well mixed in the atmosphere. Thus, only a few measuring stations distributed over both hemispheres are necessary to monitor the status and trend of tropospheric CO_2 content (Figure 4, bottom).

The most recent publications mention a value of approximately 800 billion tons for total global yearly CO_2 emissions. The uncertainties about this number are almost entirely because of the emissions from natural sources and are taken into account by specifying a range of 600 to 1,020 billion tons CO_2 (Figure 5). Manmade carbon dioxide emissions worldwide contribute only 3.5% to total global CO_2 emissions.

Power stations, home heating, and industry are the largest emitters of anthropogenic CO_2 emissions. The share of total road traffic worldwide (i.e., from passenger cars, heavy-duty vehicles, buses, and motorcycles) is approximately 13.5%. With regard to the sum of natural and anthropogenic CO_2 emissions, road traffic emissions play a minor role, contributing only about 0.5%.

Literature describes the contribution caused by the burning of biomass to be approximately 15%, but this value is highly uncertain (Figure 6). Air traffic is gaining increasing importance and has already reached the level equivalent to half the emissions from passenger car traffic.

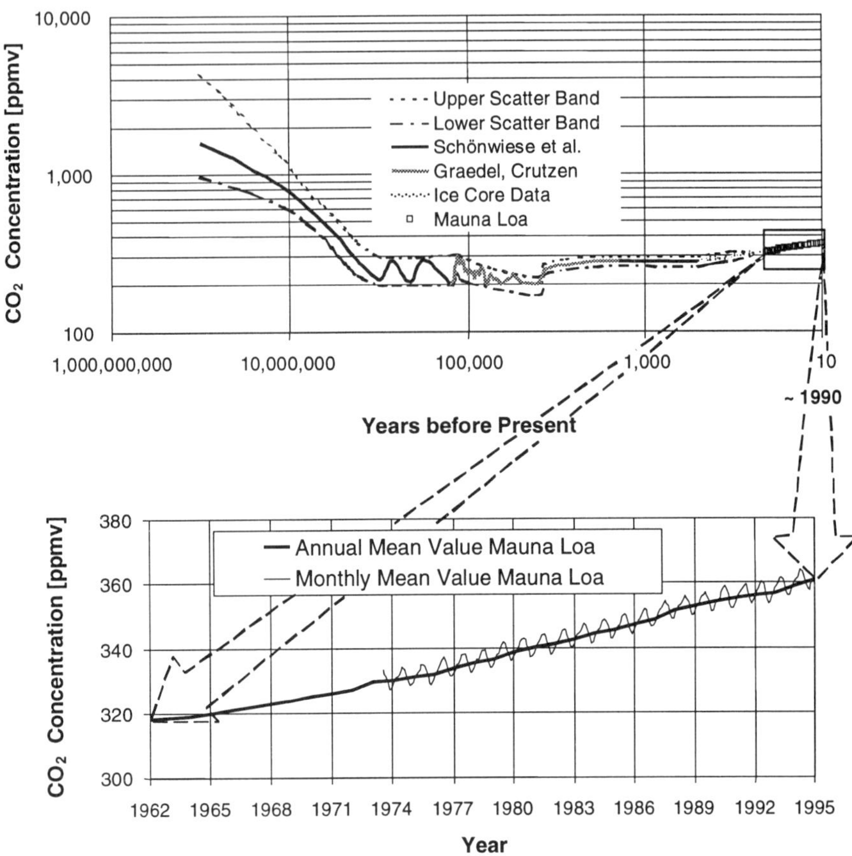

Figure 4. Development of carbon dioxide (CO_2) content in the atmosphere [21, 23].

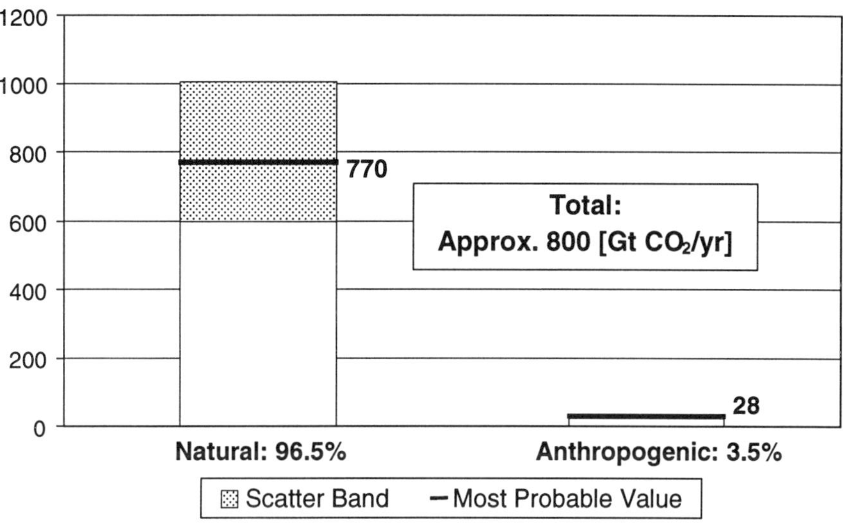

Figure 5. Spreads of global yearly carbon dioxide (CO_2) emissions, taken from various literature sources (reference year 1996) [7, 16, 24–30].

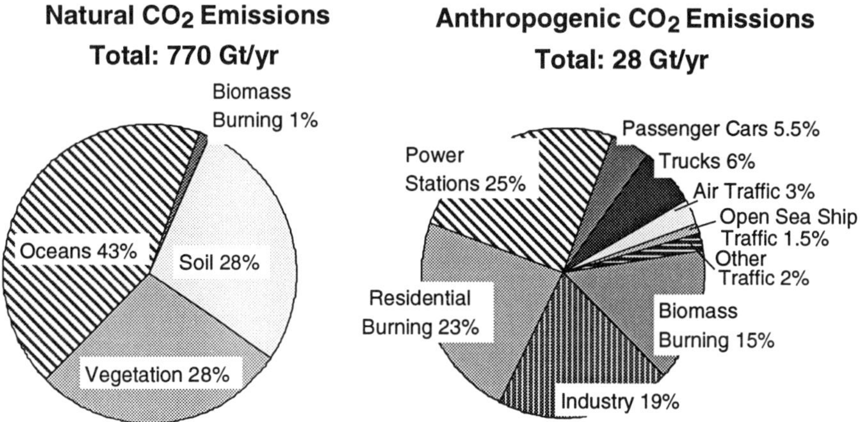

Figure 6. Contribution of various sources to global yearly carbon dioxide (CO_2) emissions, based on most probable values (reference year 1996). [For literature, see Figure 5.]

Despite efforts in many industrialized nations to reduce CO_2, these emissions will nevertheless increase in the future. Because of their increasing energy consumption in the course of economic growth, countries in Asia, Oceania, and Central and South America will contribute the greatest amount of CO_2 emissions during the coming years. However, CO_2 emissions in North America and Europe will continue to decrease in the future (Figure 7).

Figure 8 shows the development of global CO_2 emissions. The marginal share of emissions from passenger cars and road traffic is evident.

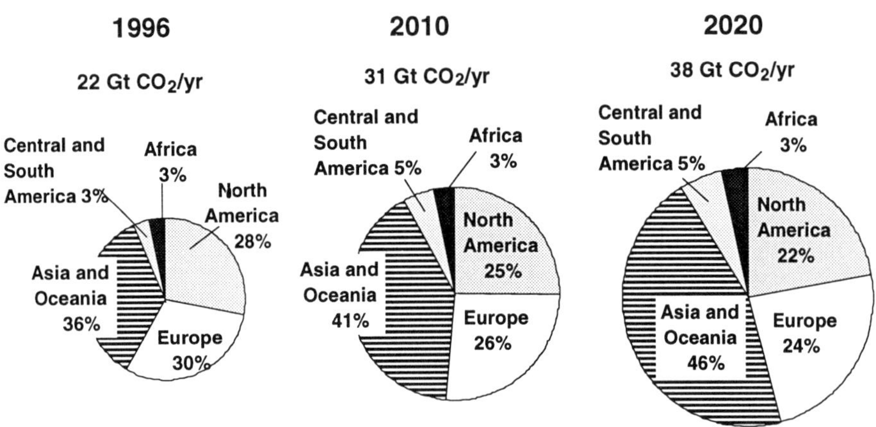

Figure 7. Prognosis for the contribution of individual continents to carbon dioxide (CO_2) emissions from fossil fuel burning [33–36].

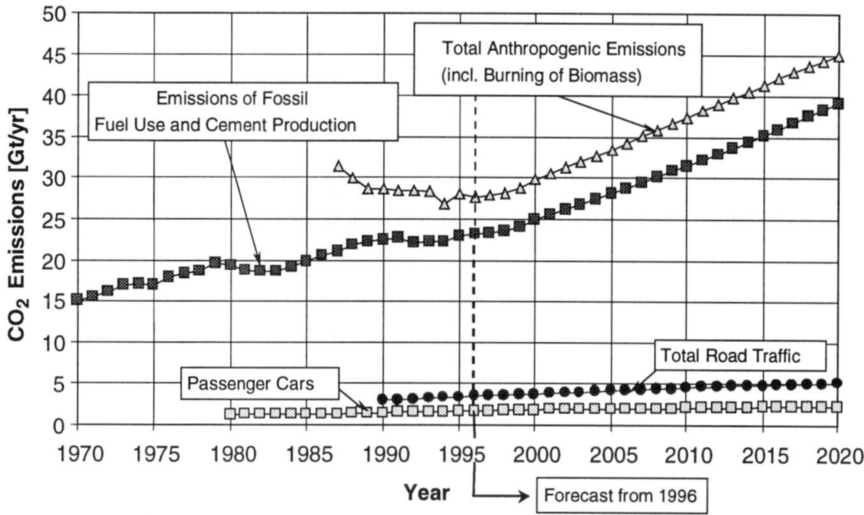

Figure 8. Trend of global anthropogenic carbon dioxide (CO_2) emissions [7, 28, 31–37].

3.2.3 Halogenated Hydrocarbons

Halogenated hydrocarbons are defined as organic compounds containing elements such as fluorine, chlorine, bromine, and iodine, which belong to the group of salt-forming substances (halogens) [38]. Halogenated hydrocarbons were developed in the 1930s, with the objective to improve comfort of lifestyle. These substances are being used as solvents, as filling for air conditioning devices and heat pumps, for the dielectric medium in transformers, as pressurized gas, in fire extinguishers, as cleansers, as pesticides, and in medications. The negative impact of these substances on the atmosphere was scientifically proven only recently, more than six decades later. Only evaporative halogenated hydrocarbons will be discussed here.

Halogenated hydrocarbons are almost exclusively of anthropogenic origin. Exceptions are methylchloride (CH_3Cl), which is released from oceans and through the burning of biomass, and partly methylbromide (CH_3Br), which is emitted by ocean algae. Partly halogenated hydrocarbons are molecules that have bound a hydrogen atom instead of a halogen atom, whereas totally halogenated hydrocarbons do not have a hydrogen atom.

Two important characteristics attributed to these hydrocarbons do not occur in nature: (1) It is assumed that they deplete the stratospheric ozone layer, and (2) They possess a strong greenhouse potential.

Any substantial short- or medium-term depletion mechanisms for totally halogenated hydrocarbons do not exist. These substances decompose only after a long time frame (which may last up to several hundred years) by some tropospheric mechanisms and through photodissociation in the middle and upper stratosphere. This decomposition in turn releases chlorine, which acts as a catalyst for ozone depletion.

The pronounced greenhouse effect inherent in halogenated hydrocarbons results from the stronger infrared absorption, which is shown by these substances in a specific part of the spectrum called the "atmospheric window." Other atmospheric gases show significantly less absorption potential within this range. Therefore, the global warming potential (GWP) of CFC 11 is, for a 20-year integration time, 4,500 times higher than that of CO_2. The GWP of CFC 12 is 7,100 times higher than that of CO_2 [39].

Partly halogenated hydrocarbons exhibit better depletion behavior. Mainly, they react with OH ions in the troposphere and decompose within a few years, thereby having a less damaging effect on the ozone layer. However, their effect on the climate for a 20-year integration time is almost as strong as that of totally halogenated hydrocarbons. Pronounced advantages for the use of partly halogenated hydrocarbons appear only when longer integration times, such as 100 or even 500 years, are considered.

In contrast to the procedure applied so far, no detailed information is available regarding the quantity of the contribution of the individual emitters because such data is unavailable in the literature. Therefore, only the total emission of the individual halogenated hydrocarbon species will be discussed (Table 3).

Figure 9 shows the global anthropogenic CO_2-equivalents of halogenated hydrocarbons for integration times of 20 and 100 years, for which we find values of 4.5 Gt/yr and 3.6 Gt/yr CO_2-equivalents, respectively.

TABLE 3. MOST PROBABLE VALUES FOR GLOBAL ANTHROPOGENIC HALOGENATED HYDROCARBON EMISSIONS FOR THE BEGINNING OF THE 1990s AND CORRESPONDING GLOBAL WARMING POTENTIAL (GWP) VALUES FOR TIME PERIODS OF 20 AND 100 YEARS [39, 41–44]

Global Halogenated Hydrocarbon Emissions at the Beginning of the 1990s	Emission [kt/yr]	GWP for 20 Years	GWP for 100 Years
CCl_3F (CFC 11)	330	4,500	3,500
CCl_2F_2 (CFC 12)	440	7,100	7,300
$C_2Cl_3F_3$ (CFC 113)	214.7	4,500	4,200
$C_2Cl_2F_4$ (CFC 114)	10.3	6,000	6,900
C_2ClF_5 (CFC 115)	12.2	5,500	6,900
$HCClF_2$ (HCFC 22)	120	4,100	1,500
$CClBrF_2$ (Halon 1211)	3	Presently no data available	Presently no data available
$CBrF_3$ (Halon 1301)	6.94	5,800	5,800
Carbon Tetrachloride CCl_4	90	1,900	1,300
1,1,1-Trichloroethane CH_3CCl_3	810	350	100
Weighted Average	–	3,290	2,870

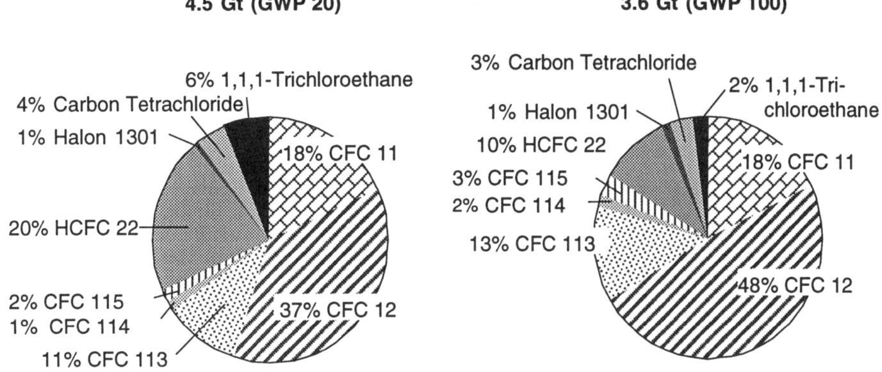

Figure 9. Global halogenated hydrocarbon emissions as CO_2-equivalents for time periods of 20 and 100 years (reference time frame mid-1990s).

The Montreal Protocol

Due to the negative effects of CFCs, the largest industrialized nations signed the Montreal Protocol on September 16, 1987. This protocol envisaged a 50% reduction of the production and use of CFCs by the year 2000. It was amended in 1990 in London and again in 1992 in Copenhagen toward commitments that were more strict. The Copenhagen amendment called for a mandatory phase-out of CFCs by 1996, a date met by the United States. Member states of the European Union committed to stop production and use of CFCs by 1995 [45, 46].

Since 1991, the automobile industry has supported the advanced withdrawal from CFC use. As an example, the production of foam parts and the filling of air conditioning systems in cars has been converted to the more environmentally friendly medium HFC 134a. For older vehicles, many car manufacturers offer a retrofit kit that allows the use of HFC 134a instead of the former chlorine-based cooling medium [146, 147].

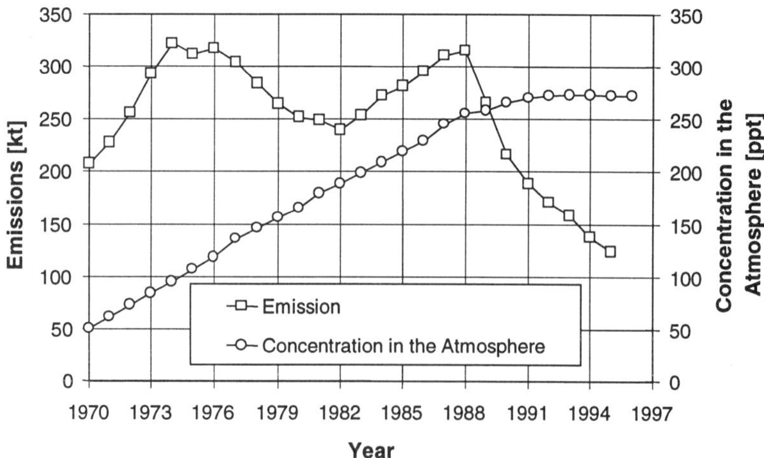

Figure 10. Trend of global emissions and concentrations in the atmosphere for CFC 11 [41, 42].

The reduction of CFC emissions after the Montreal Protocol is evident from air quality measurement results as shown in Figure 10 on the example of CFC 11. Because of the long lifetime these substances have in the atmosphere, the results do not immediately show a decreasing trend when emissions are reduced; rather, they remain at a certain level and eventually drop during subsequent years.

3.2.4 Methane

Methane results from anaerobic bacterial decay of organic substances and through thermocatalysis during crude oil formation in the surface of the earth. In particular, methane is produced during digestion from ruminants and generally under conditions in which organic material is covered by water or airtight soil layers such as those of wetlands. Wetlands can also be created by humans, as in the case of wet rice growing. Methane is further released into the atmosphere during the production and use of fossil energies [47].

Methane is depleted from the atmosphere through reaction with photolytically formed OH radicals. Because carbon monoxide is a reaction partner for OH radicals and thereby "competes" with methane, methane depletion is somewhat slowed. This effect may be the reason for the steep yearly increase in atmospheric methane concentrations of 1%, as shown in Figure 11.

Figure 12 shows the level and ranges for natural and anthropogenic methane emissions. Figure 13 summarizes the individual contributions to total global methane emissions from the various sources. Anthropogenic methane emissions, which are generated primarily from rice growing in wetlands and the breeding of ruminants, outweigh methane emissions from natural sources.

Passenger car and heavy-duty vehicle traffic contribute only a negligible 0.3%, in comparison with 0.2% to total manmade methane emissions.

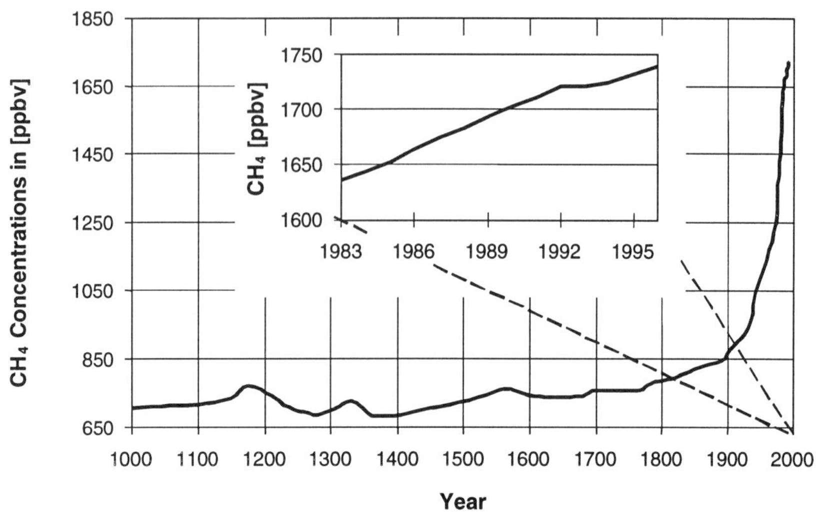

Figure 11. Development of methane (CH_4) concentrations in the atmosphere [21, 48].

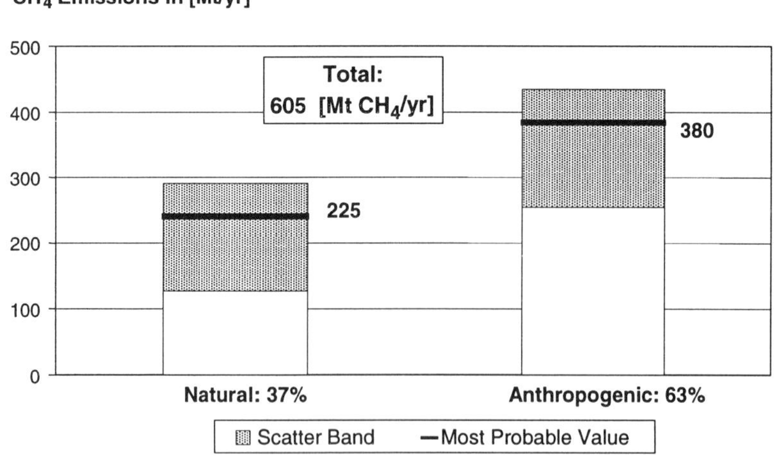

Figure 12. Spreads for global natural and anthropogenic methane (CH_4) emissions mentioned in literature (reference year 1996) [25, 32, 35, 36, 49–59].

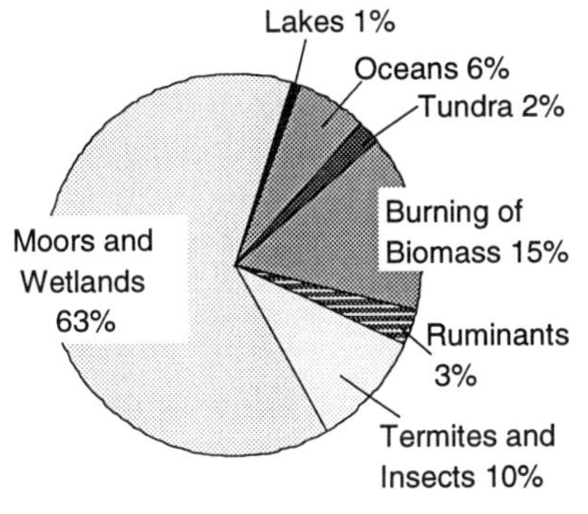

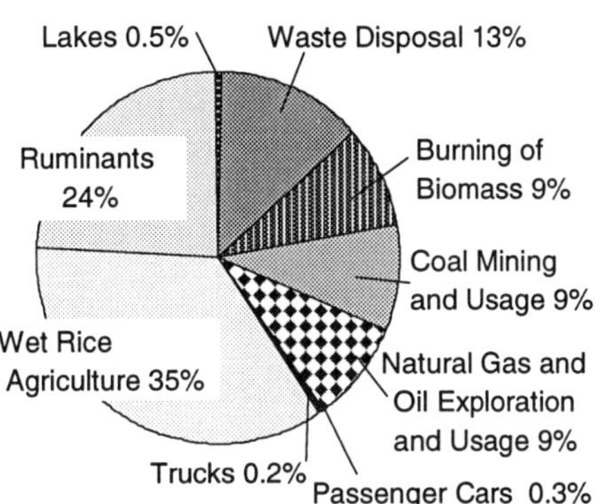

Figure 13. Contribution of various sources to global yearly methane (CH_4) emissions (reference year 1996). [For literature, see Figure 12.]

3.2.5 Nitrous Oxide

Because of its characteristics, nitrous oxide holds a special place within the group of nitrogen oxides. The term "nitrogen oxides" refers to nitrogen compounds that may exist in various forms [47]. The most important ones are as follows:

- Nitric oxide (NO)
- Nitrogen dioxide (NO_2)
- Nitrogen trioxide (NO_3)
- Nitrous oxide (N_2O) and other compounds such as N_2O_3, N_2O_4, and N_2O_5

Nitrous oxide is a colorless, odorless gas that is soluble in water. In 1938, it became possible to trace nitrous oxide in the atmosphere by measuring its infrared absorption at 7.57 and 7.77 μm. These two values are found in that part of the spectrum known as the "atmospheric window" because other atmospheric gases show only limited absorption capability in this area. This explains why N_2O causes such a strong greenhouse effect [60]. Also, N_2O absorbs 200 times as much as CO_2 in the wavelength range that is responsible for this effect [61]. In Reference 12, this value is even mentioned as a factor of 290 at an integration time of more than 100 years.

Tropospheric N_2O is being formed by nitrifying microorganisms in soil and sea water, as well as in the groundwater of the continents. Nitrous oxide is chemically inert in the troposphere, whereas it will be photolytically decomposed or will react with atomic oxygen in the stratosphere. This reaction generates molecular nitrogen and nitric oxide (NO). Nitrous oxide seems to be the primary source for stratospheric NO and therefore is considered to participate in the ozone depletion process also via this route [61].

Due to its long lifetime, N_2O is homogeneously distributed in the atmosphere. Although small variations occur from year to year, a distinct yearly cycle of tropospheric N_2O cannot be detected. As of 1990, the average global N_2O concentration in the atmosphere reached a level of 310 ppbv (Figure 14).

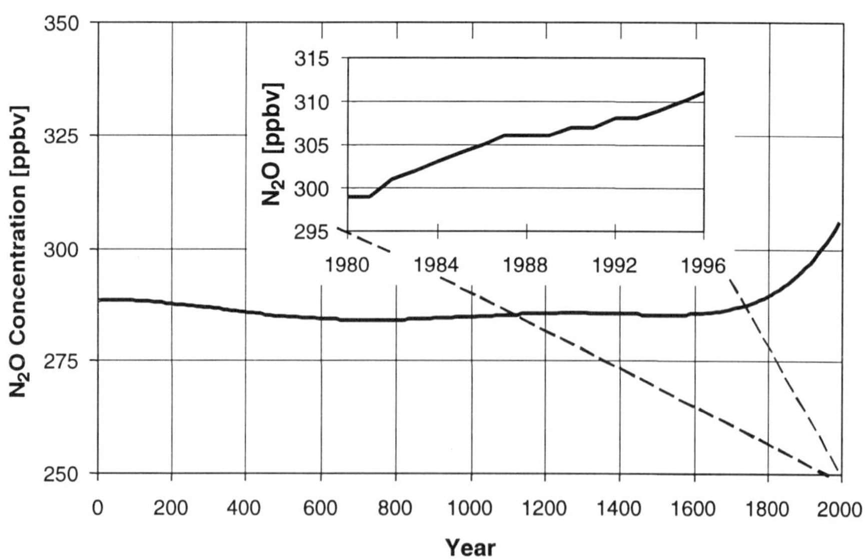

Figure 14. Development of nitrous oxide (N_2O) concentrations in the atmosphere [21, 48].

The primary anthropogenic N_2O emission source is agriculture. Excess amounts of fertilizer that have not been washed out normally decompose through denitrification. Generally, all combustion processes can represent N_2O emission sources. In addition, N_2O losses into the atmosphere occur during its production and use and during waste utilization.

Figure 15 shows the distribution between natural and anthropogenic sources with respect to total global N_2O emissions, together with the corresponding ranges. Figure 16 shows the contributions of the individual emitters.

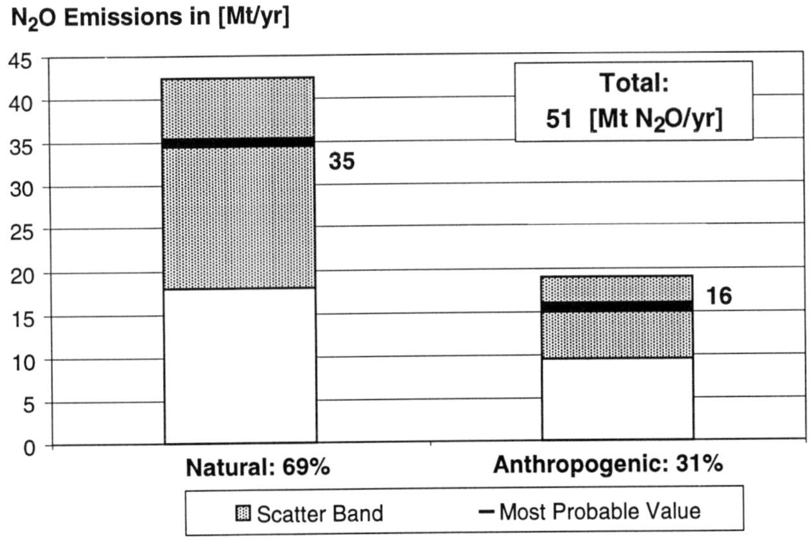

Figure 15. Spreads of global natural and anthropogenic nitrous oxide (N_2O) emissions discussed in the literature [25, 61–66].

The major sources of natural N_2O emissions are the oceans, followed by tropical forests. Anthropogenic emissions that stem mainly from the use of fertilizers in agriculture, followed by slash-and-burn of tropical forests and industrial processes, represent only half the natural emissions.

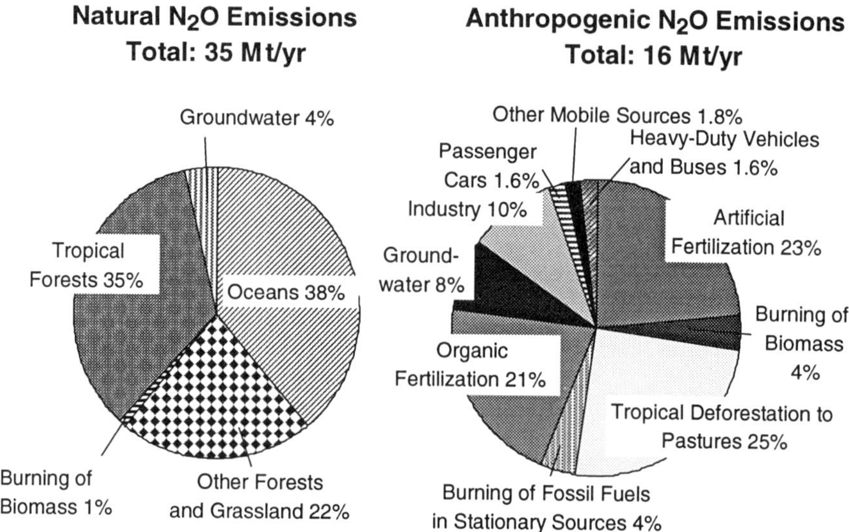

Figure 16. Contribution of various sources to global yearly nitrous oxide (N_2O) emissions based on most probable values (reference year 1996). [For literature, see Figure 15.]

Industrial processes are considered to include the use of N_2O for medical purposes (anesthesia). It was impossible to determine the amount of worldwide N_2O emissions resulting from this use. However, an estimate is given in Reference 133 for the Federal Republic of Germany, attributing one-third of the emissions caused by the use of N_2O to medical applications.

Passenger car and heavy-duty vehicle traffic each contribute 1.6% to total global anthropogenic nitrous gas emissions. Hence, they play only a subordinate role.

In spite of this marginal contribution of road traffic, attention recently has been drawn to the increasing number of passenger cars having gasoline engines and three-way catalysts that generate nitrous gas emissions as an undesirable by-product of NO_x conversion. According to current scientific knowledge, the mechanisms of N_2O formation in the three-way catalyst are understood. Therefore, we can anticipate that these N_2O emissions will decrease as a result of improved catalyst technology and the corresponding governing methods for engine operation [81].

CHAPTER 4

Exhaust Emission Components with Local and Regional Effects

Emissions of exhaust components having a short life span can cause an increase in local concentrations of individual substances and thus may lead to local and regional effects such as smog and acid rain.

Smog

The term "smog" originates from an episode in London at the end of the 1800s when a mixture of smoke and fog choked the city. This episode was caused by high emissions of sulfur dioxide (SO_2) and soot from the burning of coal with high sulfur content. Although soot promotes the formation of fog (which always occurs, especially during winter), sulfur dioxide dissolved in the fog droplets and then formed sulfurous acid and sulfuric acid. This kind of smog is referred to in literature as "London smog." Smog disasters are not expected to occur in Western economies today because of improved air quality achieved through effective control measures (see Section 4.3 on sulfur dioxide) [67].

In contrast to the London situation, smog also can form through photochemical processes in the atmosphere of fog-free congested areas. Photochemical smog is generated when nitrogen oxides, carbon monoxide, and hydrocarbons are subjected to radiation from intense sunlight. Such a situation creates highly oxidizing substances, especially ozone, as well as airborne components that create haze and reduce visibility. This phenomenon first occurred in Los Angeles, where only limited air exchange is possible due to the unique geographical situation, and thus it is referred to as "Los Angeles smog." Due to the potentially negative effects of smog episodes on health, air quality protection laws have been established to reduce smog precursors.

Acid Rain and Damage to Forests

During the first half of the 1800s, it became known that rain may contain acids that can negatively impact vegetation by accelerating soil acidification and directly attacking leaves [3, 67, 68].

Today, it is believed that the negative effects on trees result from the combined impact of manifold anthropogenic and natural environmental factors. These factors can, among other things, result from climatic influences (i.e., frost and drought periods), unfavorable conditions of location and soil constitution, viruses, air pollution, over-fertilization, over-aged forest stock, and organisms from flora and fauna. Many changes to trees and forests occur within their natural life cycles, without any anthropogenic influence.

The most recent reports on the status of forests in Europe show that the so-called "forest damage" has decreased [91], and an increase of 43% in wood generation (trees, bushes, shrubs, etc.) was realized between 1950 and 1990 [92].

4.1 Nitrogen Oxides

For this discussion on the aspects of air quality and photochemical reactions in the atmosphere, two gases (namely, NO and N_2O) are most important. Mixtures of these gases fall under the term nitrogen oxides, abbreviated as "NO_x," which means Σ (NO, N_2O). NO_x emissions (calculated emissions) will always be given here as NO_2.

Nitrogen oxides (NO_x) are formed, for example, during combustion processes with high temperatures. The higher the efficiency of the combustion process (i.e., the lower the corresponding generation of CO_2), the more NO_x is formed. This interrelationship creates a physical conflict with efforts toward the reduction of CO_2 emissions (Figure 17).

Anthropogenic sources of NO_x emissions are road traffic, industry, power plants, home heating, small utilities, waste incineration, and agriculture. Emissions from agriculture stem primarily from the use of nitrogen-containing fertilizer. However, it must be mentioned that nitric oxide evaporates from soil in small amounts even without the use of such fertilizer.

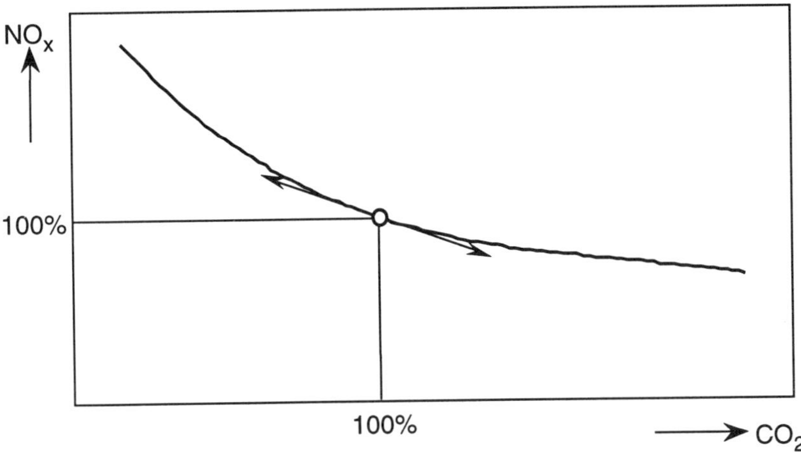

Figure 17. Basic conflict between NO_x and CO_2 emissions for an engine combustion process [127].

Analogous to the previous progress report [127] and due to the importance attributed to nitrogen oxides in the framework of European efforts to improve air quality, the global NO_x emission balance will be highlighted here again.

In contrast to regional considerations, the contributions from natural sources are taken into account within global emission balancing. Such evaluation reveals that natural emissions approximately equal the contribution caused by anthropogenic sources (Figure 18). These natural emissions stem primarily from soil and lightning (Figure 19).

Passenger car and heavy-duty vehicle traffic contribute 20% to total anthropogenic emissions worldwide.

4.1.1 Nitrogen Oxide Emissions in Europe

Emissions of nitrogen oxides in Europe reached a culmination point in 1989 (Figure 20).

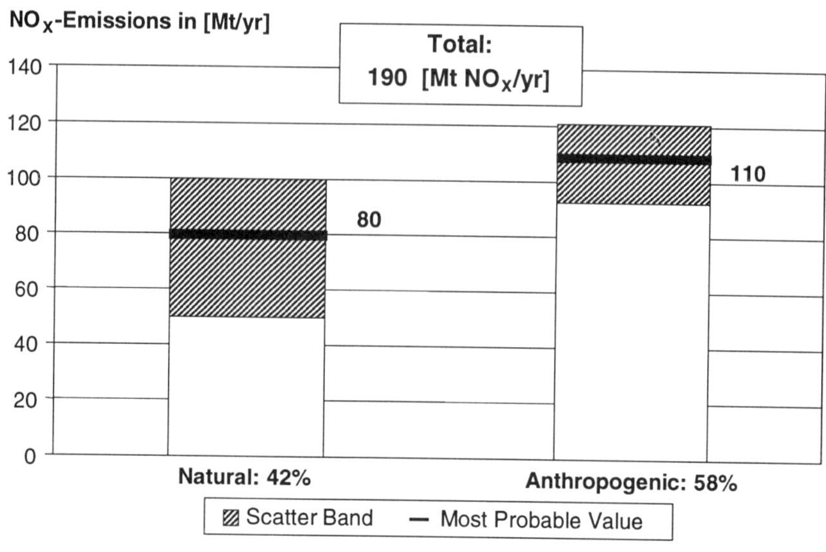

Figure 18. Spreads of global yearly nitrogen oxide (NO_x) emissions from various literature sources [5, 36, 49 , 53, 61, 63, 75, 76, 110, 140–142].

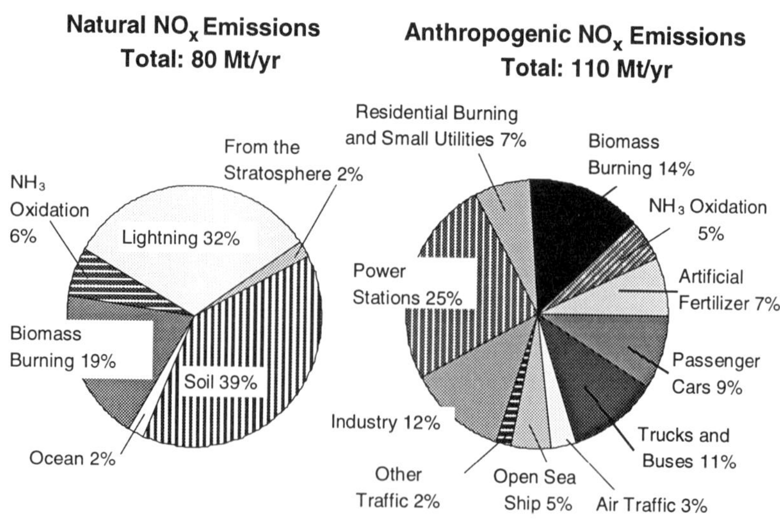

Figure 19. Contribution of various sources to global yearly nitrogen oxide (NO_x) emissions based on most probable values (reference year 1996). [For literature, see Figure 18.]

Road traffic in Europe contributes approximately 50% to nitrogen oxide emissions (Figure 21). One literature source [74] calculates the share of agriculture as 22%, but other authors [25] estimate this value to 35%.

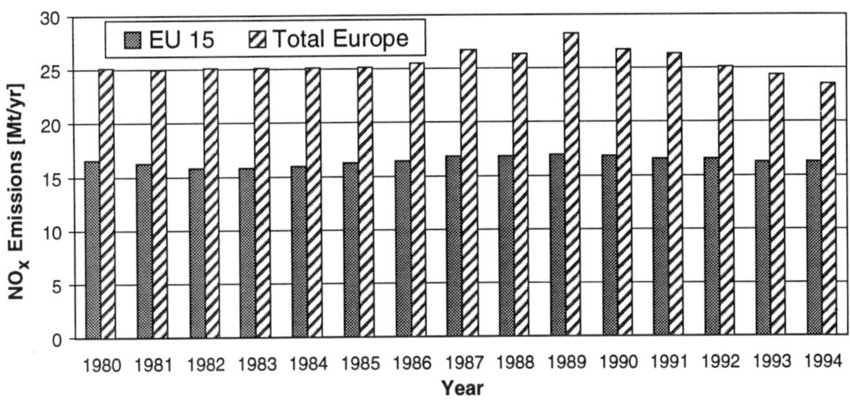

Figure 20. Trend of nitrogen oxide (NO_x) emissions in Europe and in the European Union—15 States (EU15) [25, 69–74].

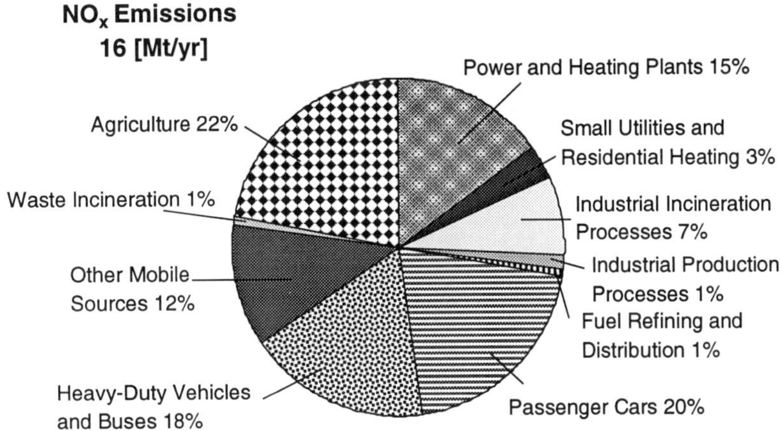

Figure 21. Contribution of various sources to nitrogen oxide (NO_x) emissions in the European Union—15 States (EU15) (reference year 1994). [For literature, see Figure 20.]

Nitrogen oxide emissions caused by natural events such as lightning contribute approximately 13% to total emissions worldwide. However, a comparable estimate for Europe could not be found in the literature.

4.1.2 Nitrogen Oxide Emissions in the Federal Republic of Germany

Until 1988, power and heating plants were the primary emitters of NO_x emissions. However, due to conversion to low-emission heating systems, de-NO_x measures, and the closing of high-emitting plants, emissions from these sectors dropped drastically (Figure 22). In spite of the substantial increase between 1970 and 1996 in passenger cars from 14 to 41 million vehicles, NO_x emissions from this car segment have decreased continuously by more than 30% because of the introduction of effective emission control systems. Figure 23 shows the contributions of individual sources to NO_x emissions in the Federal Republic of Germany.

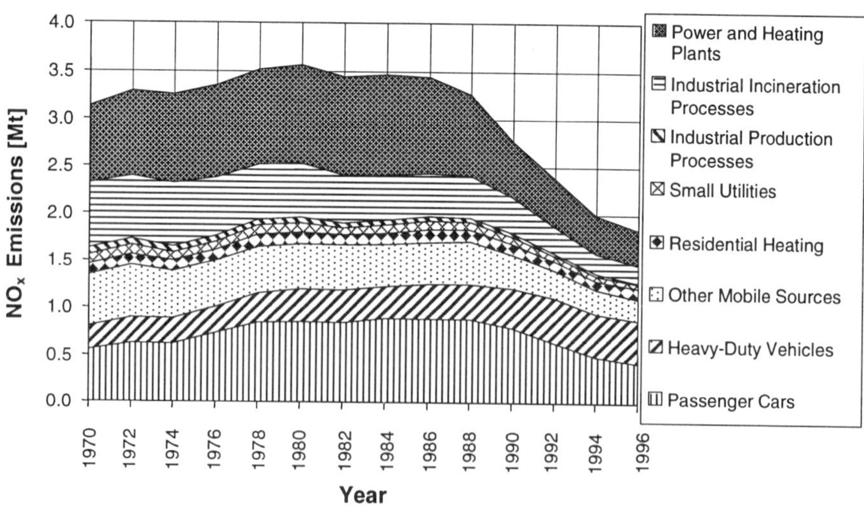

Figure 22. Trend of nitrogen oxide (NO_x) emissions in the Federal Republic of Germany [69, personal calculations].

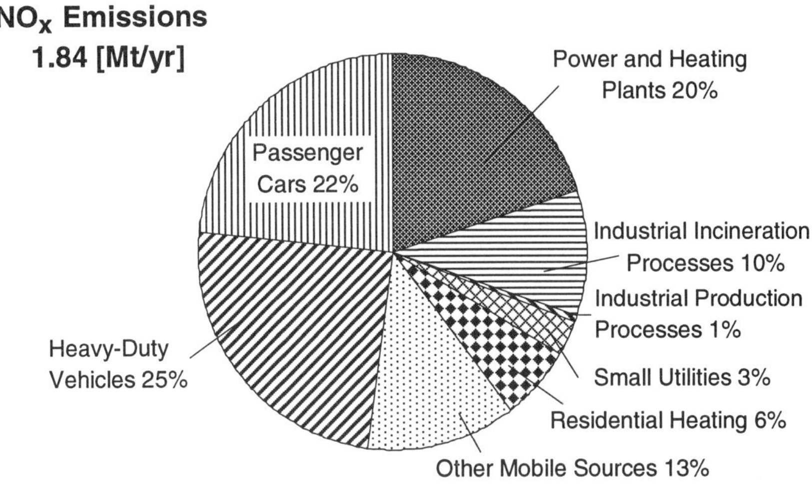

Figure 23. Contribution of various sources to nitrogen oxide (NO_x) emissions in the Federal Republic of Germany (reference year 1996) [69, personal calculations].

4.1.3 Judgment Criteria for the Impact of Nitrogen Oxides on Air Quality

As judgment criteria for the impact of nitrogen oxides on air quality, guide and limit values have been specified in the "Technische Anleitung zur Reinhaltung der Luft"—TA Luft (Technical Guideline for the Purification of the Air), in the "23. Verordnung zum Bundes-Immissions-Schutzgesetz" (23rd Ordinance to the Federal Air Quality Protection Law), in the "Richtlinie 2310 des Vereins Deutscher Ingenieure" (Guideline 2310 of the Association of German Engineers), in the EEC-Directive 85/203/EEC, and in the Guidelines of the World Health Organization (WHO) (Tables 4 and 5).

TABLE 4. LIMIT AND GUIDE VALUES FOR NITROGEN DIOXIDE (NO_2) CONCENTRATIONS IN THE ATMOSPHERE

EU-Directive 85/203/EEC Appendix I			
Concentration Value	**Reference Time Frame**	**Protection Objective**	**Value Character**
200 μg/m³	98%-Value of Sum Frequency of Yearly One-Hour Average	Human Health	Limit Value
EU-Directive 85/203/EEC Appendix II			
Concentration Value	**Reference Time Frame**	**Protection Objective**	**Value Character**
50 μg/m³	Median Value of Yearly One-Hour Average	Human Health	Guide Value
135 μg/m³	98%-Value of Sum Frequency of Yearly One-Hour Average	Human Health	Guide Value
23rd Ordinance to the Federal Air Quality Protection Law			
Concentration Value	**Reference Time Frame**	**Protection Objective**	**Value Character**
160 μg/m³	98%-Value of Sum Frequency of Yearly One-Hour Average	Human Health	Guide Value
Technical Guideline—Air			
Concentration Value	**Reference Time Frame**	**Protection Objective**	**Value Character**
80 μg/m³	Arithmetic Yearly Average of Half-Hour Average Values	Human Health	Limit Value
200 μg/m³	98%-Value of Sum Frequency of Yearly Half-Hour Average	Human Health	Limit Value
WHO			
Concentration Value	**Reference Time Frame**	**Protection Objective**	**Value Character**
40 μg/m³	Yearly Average	Human Health	Guide Value
200 μg/m³	Daily Average	Human Health	Guide Value
VDI-Guideline 2310 (MIK-Value)			
Concentration Value	**Reference Time Frame**	**Protection Objective**	**Value Character**
100 μg/m³	Daily Average	Human Health	Guide Value
200 μg/m³	Half-Hour Average	Human Health	Guide Value

TABLE 5. LIMIT AND GUIDE VALUES FOR NITRIC OXIDE (NO) CONCENTRATIONS IN THE ATMOSPHERE

VDI-Guideline 2310			
Concentration Value	**Reference Time Frame**	**Protection Objective**	**Value Character**
500 µg/m³	Daily Average	Human Health	Guide Value
1000 µg/m³	Half-Hour Average	Human Health	Guide Value

4.1.4 Concentrations of Nitrogen Oxides in the Atmosphere

Figure 24 shows the development of nitrogen oxide concentrations in the atmosphere at selected measuring stations in the Federal Republic of Germany and Austria. The concentrations are decreasing continuously at the urban background and at the roadside stations.

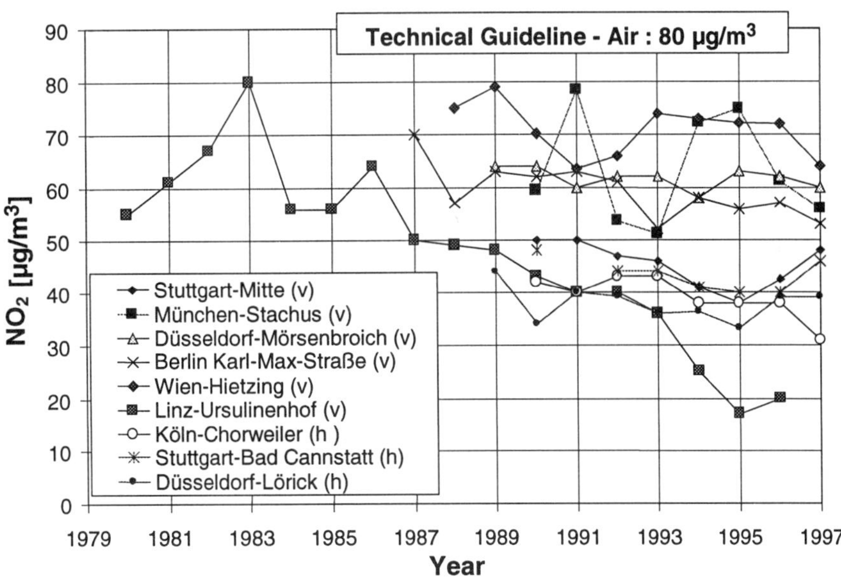

Figure 24. Development of yearly average values for nitrogen dioxide (NO_2) concentrations in the atmosphere at roadside measuring stations (v) and urban background measuring stations (h) in the Federal Republic of Germany and Austria [78–80, 83, 89, 131].

Measurements at clean air locations in the Federal Republic of Germany show, independently from emission reduction measures for traffic and industry, an approximately constant level over two decades (Figure 25).

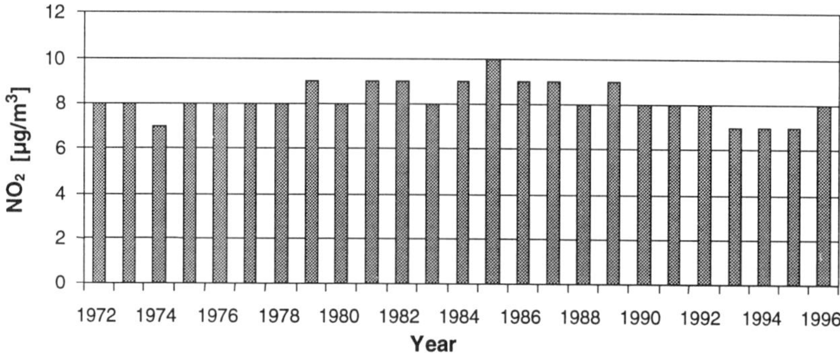

Figure 25. Development of yearly average values for nitrogen oxides at the rural measuring stations of the Federal Ministry of Germany [69].

4.2 Non-Methane Hydrocarbons

"Classic" components such as carbon monoxide, sulfur dioxide, and nitrogen oxides are clearly defined. However, the term "hydrocarbons" comprises a mixture of a large variety of compounds. The term is generally understood to cover not only pure hydrocarbons but also other organic compounds. Nevertheless, uncertainties may occur from time to time about which chemical class of substances must be included [81].

In addition, "organic compounds" include non-volatile substances that normally exist in liquid or solid chemical compounds and thus cannot represent "air" contaminants. Thus, emission values are generally related to volatile organic compounds (VOC).

Methane is a special case and therefore is excluded when emission values are given as the sum of organic compounds without methane (NMVOC). This study uses the term "non-methane hydrocarbons" (NMHC), which means the same as "non-methane volatile organic compounds" (NMVOC).

4.2.1 Non-Methane Hydrocarbon Emissions in Europe

Non-methane hydrocarbon emissions in Europe follow the same trend as the nitrogen oxide emissions, with highest levels occurring around 1989 and declining since then (Figure 26).

In the reference year 1994, the highest amounts of anthropogenic NMHC emissions in the European Union resulted from the use of solvents (29%), followed by exhaust emissions from passenger cars (22%) (Figure 27).

When compared to total NMHC emissions generated in the European Union, the natural emission source forests must be considered as a major contributor that is responsible for a share of 16 to 32% [25, 71, 73].

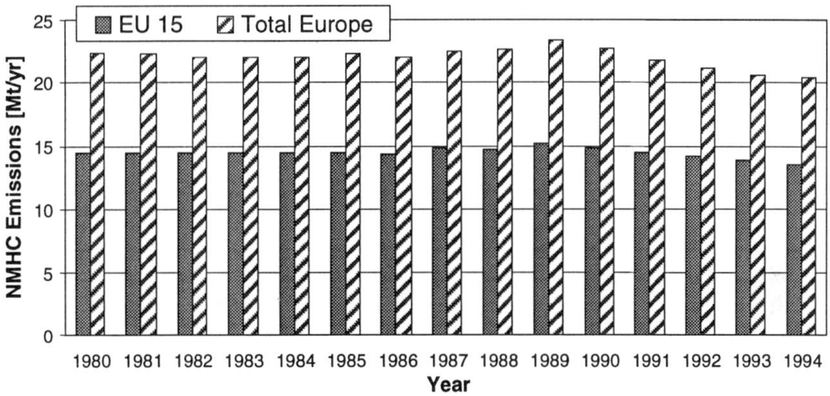

Figure 26. Development of volatile non-methane hydrocarbon (NMHC) emissions in Europe and the European Union—15 States (EU 15) [25, 70, 71, 73].

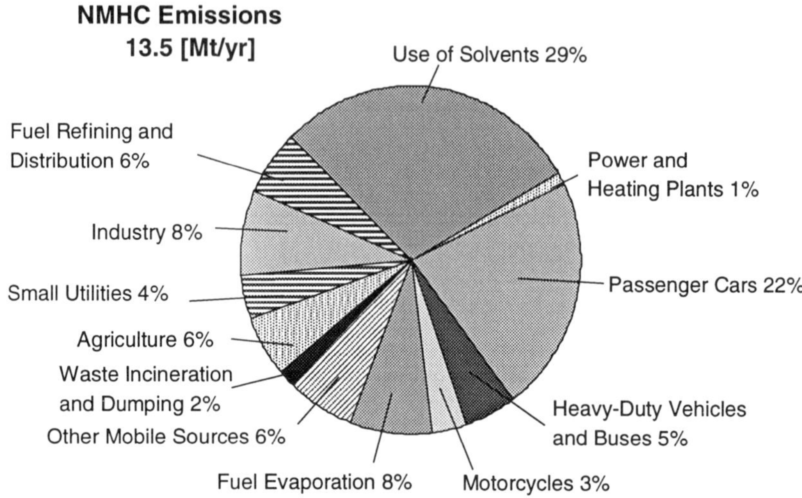

Figure 27. Contribution of various sources to yearly emissions of volatile non-methane hydrocarbons (NMHC) in the European Union—15 States (reference year 1994). [For literature, see Figure 26.]

4.2.2 Non-Methane Hydrocarbon Emissions in the Federal Republic of Germany

At the beginning of the 1970s, passenger car traffic and the use of solvents were considered the main sources of NMHC emissions in the Federal Republic of Germany (Figure 28). Through the introduction of modern engine and emission control technology, emissions from passenger cars and heavy-duty vehicles have decreased continuously since 1987. Today, evaporation from the use of solvents has reached the same level as total road traffic, the emissions of which will further lose importance because of more stringent upcoming regulations (Figure 29).

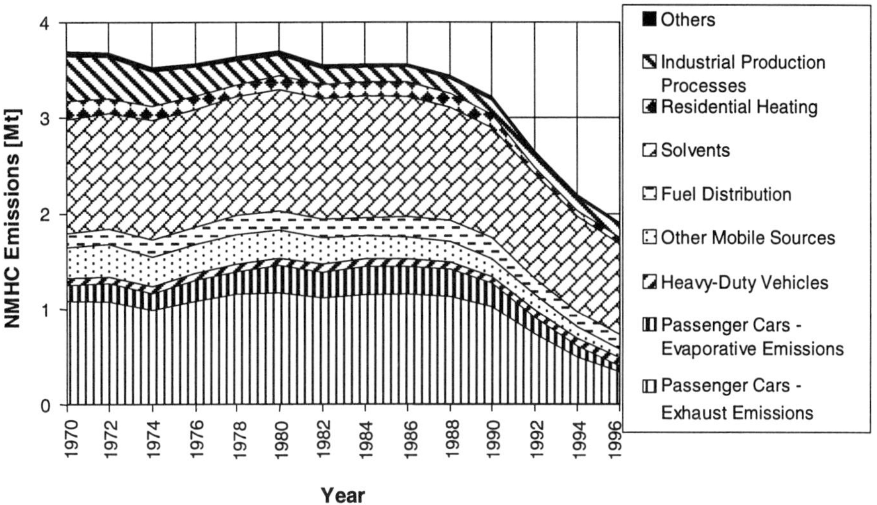

Figure 28. Trend for non-methane hydrocarbon (NMHC) emissions in the Federal Republic of Germany [69, personal calculations].

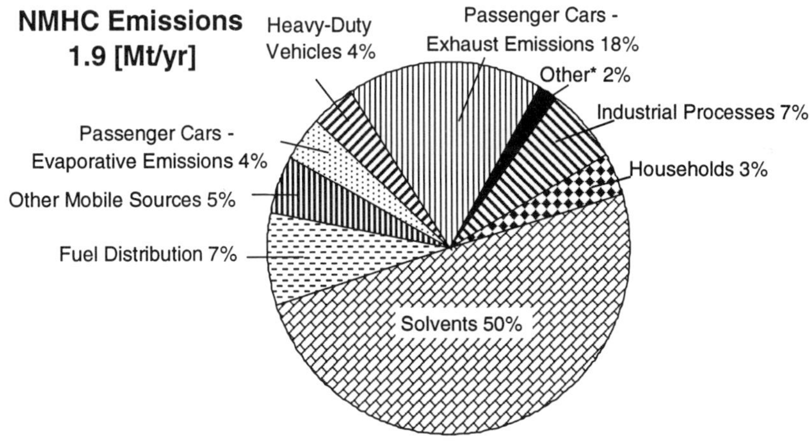

Figure 29. Contribution of various sources to non-methane hydrocarbon (NMHC) emissions in the Federal Republic of Germany (reference year 1996) [69, personal calculations].

4.2.3 Concentrations of Non-Methane Hydrocarbons in the Atmosphere

Neither limits nor guide values exist for NMHC for the characterization of air quality. A substantial improvement in air quality has occurred since the introduction of the three-way catalyst on passenger cars. The concentration of NMHC in the atmosphere dropped to one-quarter of the previous value within a period of nine years (Figure 30). Due to low concentrations in the atmosphere and the high cost of maintaining a measurement station, measurements will be stopped successively at the few stations that allow for a trend determination.

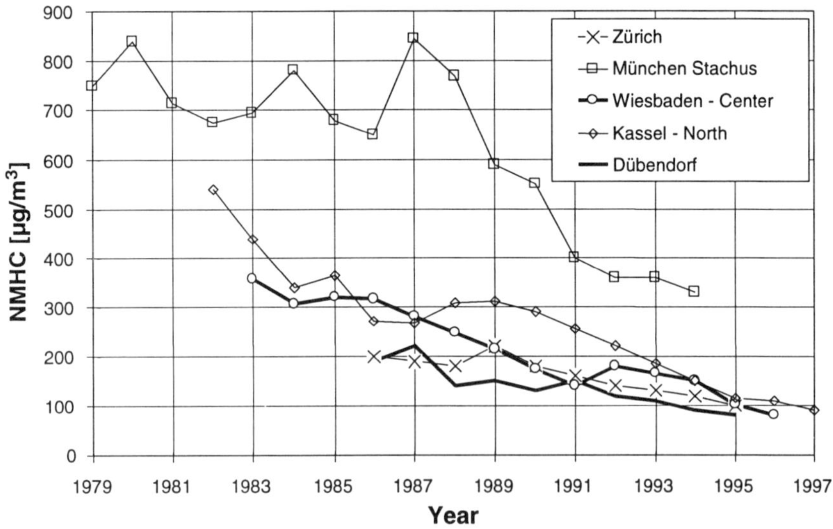

Figure 30. Development of yearly average values for non-methane hydrocarbon (NMHC) concentrations in the atmosphere at roadside measuring stations in the Federal Republic of Germany and in Switzerland [77, 82, 83].

4.3 Sulfur Dioxide

In the atmosphere, sulfur occurs in the form of oxides and sulfates. Among the oxides described in the literature, only sulfur dioxide (SO_2)—and possibly sulfur trioxide (SO_3) as a precursor of sulfate—and sulfuric acid aerosols are important. This discussion will consider only sulfur dioxide (SO_2) emissions.

Sulfur dioxide results primarily from incineration processes by oxidation of sulfur contained in fuel. Process-related emissions may result from crude oil and natural gas, as well as from metal processing and from processes in the chemical industry.

Natural sources of atmospheric sulfur are considered in two groups: biogenic and non-biogenic sources. Emissions from non-biogenic sources are volcanic activities and the burning of biomass. Biogenic sources are emissions from swamps, oceans, and tropical forests.

4.3.1 Sulfur Dioxide Emissions in Europe

Sulfur dioxide emissions are decreasing steadily in Europe, as shown in Figure 31.

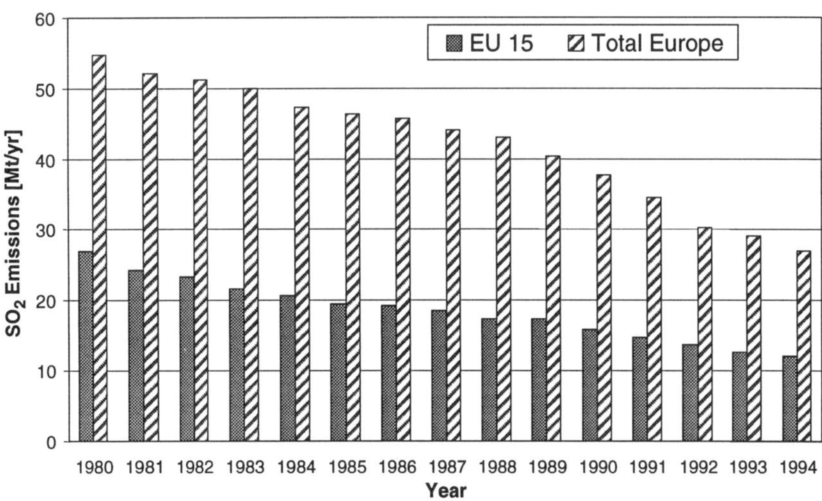

Figure 31. Trend for sulfur dioxide (SO_2) emissions in Europe and in the European Union—15 States (EU 15) [70, 71].

Among the various sources, power plants and heating plants contribute the most to total SO_2 emissions, as shown in Figure 32. Volcanoes and forest fires are mentioned in the literature as the second largest SO_2 emission source. During reference year 1994, emissions from natural sources represented 21% of total SO_2 emissions in the European Union, where emissions from forest fires cannot be separated precisely into those of anthropogenic or natural origin.

Road traffic plays only a negligible role, and its contribution will decrease further with the envisaged lowering of sulfur content in fuel.

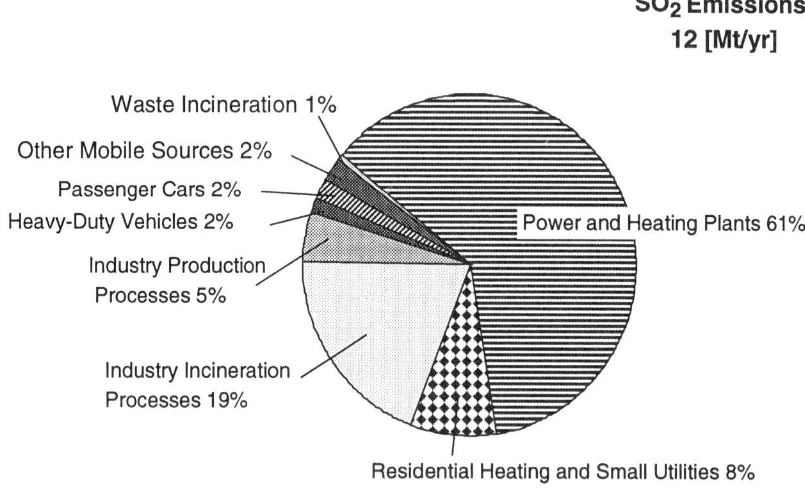

Figure 32. Contribution of various sources to yearly sulfur dioxide (SO_2) emissions in the European Union—15 States (reference year 1994) [70, 71].

4.3.2 Sulfur Dioxide Emissions in the Federal Republic of Germany

Since 1986, a substantial decline in SO_2 emissions of approximately 60% has occurred (Figure 33). This was caused by the introduction of the German "Technical Guideline—Air" and by the closing of firms in the "new" states of the Federal Republic of Germany; decreasing energy demand caused by the restructured economy; the use of low-emission fuels for energy generation; and changed consumer behavior in the population.

Power and heating plants contribute 62%, the largest share of SO_2 emissions. Due to its very low share of less than 3%, the contribution of road traffic to total SO_2 emissions can be disregarded (Figure 34).

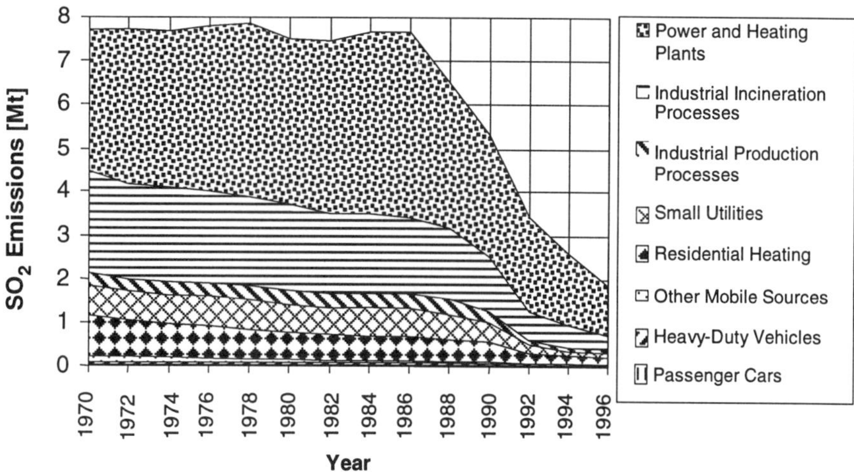

Figure 33. Trend for sulfur dioxide (SO_2) emissions in the Federal Republic of Germany. (Today's contribution of passenger car and heavy-duty vehicle traffic is so small that it cannot be shown within the scale of this graph [69, personal calculations].

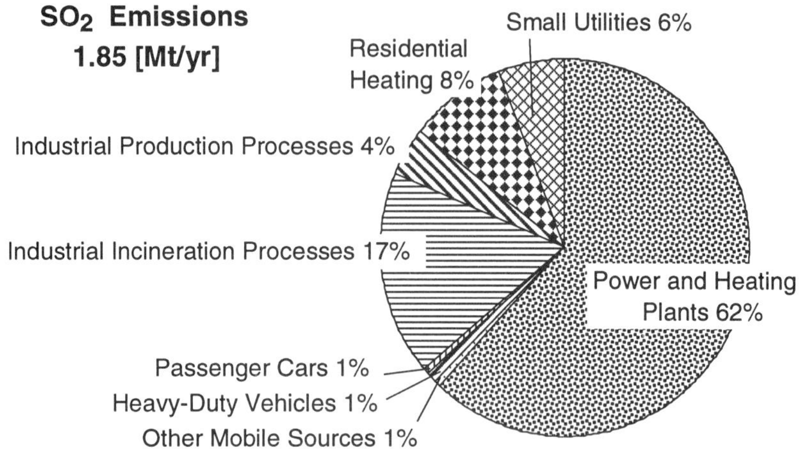

Figure 34. Contribution of various sources to sulfur dioxide (SO_2) emissions in the Federal Republic of Germany (reference year 1996) [69, personal calculations].

4.3.3 Judgment Criteria for the Impact of Sulfur Dioxide on Air Quality

The Directive of the European Union 80/779/EEC specifies in its Appendix IV values that are mentioned as limit values in the 22nd Federal Air Quality Protection Ordinance (BImSchV). This Ordinance transposes the EEC-Directive into national German law. These limit values are valid in combination with the concentration of airborne dust. Appendix II of the Ordinance specifies two additional guide values.

In Germany, the arithmetic yearly average of all half-hour values of one year (used to describe long-term effects) and their 98% sum frequency value (used to describe short-term effects) are the measures used for air quality judgment (Table 6).

Guide values for the protection of human health also are specified by the World Health Organization (WHO). These values correspond approximately to the guide values of Appendix II of the EU-Directive.

TABLE 6. GUIDE AND LIMIT VALUES FOR SULFUR DIOXIDE (SO_2) CONCENTRATIONS IN THE ATMOSPHERE IN THE FEDERAL REPUBLIC OF GERMANY

EU-Directive 80/779/EEC				
Appendices II/IV	**Concentration Value**	**Reference Time Frame**	**Protection Objective**	**Character of Value**
Appendix IV	Same as 22nd Federal Air Quality Protection Regulation		Human Health	Limit Value
Appendix II	40–60 µg/m^3	Arithmetic Average Value	Human Health and Environment	Guide Value
	100–150 µg/m^3	Daily Average Value	Human Health and Environment	Guide Value
22nd Federal Air Quality Protection Regulation				
Concentration Value	**Reference Time Frame**		**Protection Objective**	**Character of Value**
80 µg/m^3	Median of Daily Average Values Measured Over One Year If Median of Suspended Particulate Matter Concentration Is >150 µg/m^3		Human Health	Limit Value
120 µg/m^3	Median of Daily Average Values Measured Over One Year If Median of Suspended Particulate Matter Concentration Is <150 µg/m^3		Human Health	Limit Value
130 µg/m^3	Median of Daily Average Values Measured During Winter Season If Median of Suspended Particulate Matter Concentration Is >200 µg/m^3		Human Health	Limit Value
250 µg/m^3	98%-Value of Sum Frequency of Daily Average Values Measured Over One Year If 98%-Value of Suspended Particulate Matter Concentration Is >350 µg/m^3		Human Health	Limit Value
350 µg/m^3	98%-Value of Sum Frequency of Daily Average Values Measured Over One Year If 98%-Value of Suspended Particulate Matter Concentration Is <350 µg/m^3		Human Health	Limit Value

TABLE 6. (cont.)

Technical Guideline—Air			
Concentration Value	**Reference Time Frame**	**Protection Objective**	**Character of Value**
140 μg/m^3	Arithmetic Yearly Average from Half-Hour Average Values	Human Health	Limit Value
400 μg/m^3	98%-Value of Sum Frequency of Yearly Half-Hour Average Values	Human Health	Limit Value
WHO			
Concentration Value	**Reference Time Frame**	**Protection Objective**	**Character of Value**
50 μg/m^3	Yearly Average Value	Human Health	Guide Value
125 μg/m^3	Daily Average Value	Human Health	Guide Value

4.3.4 Concentrations of Sulfur Dioxide in the Atmosphere

The decrease of emissions is reflected in ambient air concentration measurement results obtained from roadside and urban background measurement stations (Figure 35). However, it must be emphasized that ambient air sulfur dioxide concentrations at roadside measurement stations are not determined by road traffic but rather are being caused by heating processes in the surrounding buildings.

The impact of SO_2 reduction measures in congested areas also results in air quality improvements in rural areas (Figure 36).

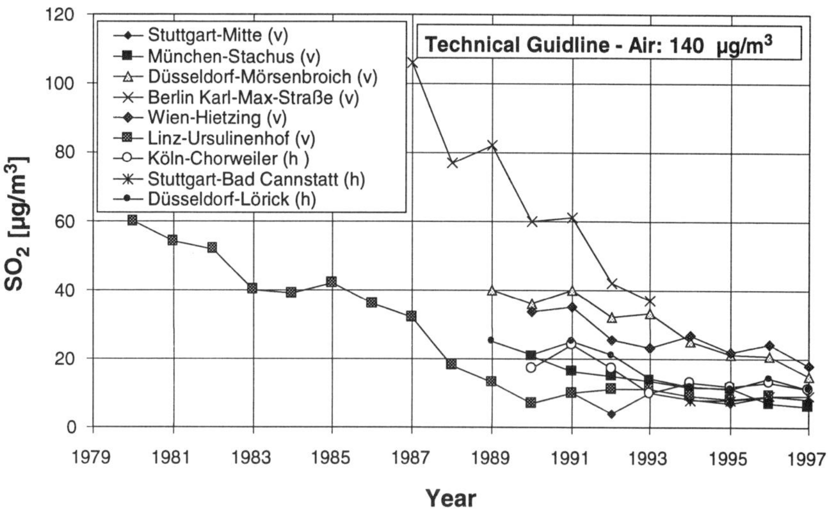

Figure 35. Development of yearly average values for sulfur dioxide (SO_2) concentrations in the atmosphere at roadside measuring stations (v) and urban background measuring stations (h) in the Federal Republic of Germany and Austria [78–80, 83, 89, 131].

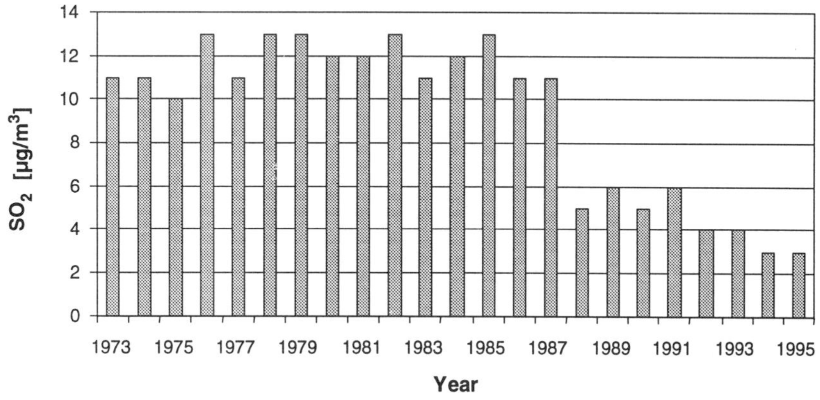

Figure 36. Development of yearly average values for sulfur dioxide (SO_2) concentrations at rural measuring stations of the Federal Ministry of Germany [69].

4.4 Dust and Particulate Matter

The terms "airborne dust" or "particulate matter" refer to airborne particles that may either be generated and emitted directly by natural or anthropogenic sources, or those that are caused indirectly from gaseous precursors and may also be of natural or anthropogenic origin [40].

Particulate emissions from direct or primary sources are as follows:

- Airborne dust
- Soot
- Smoke

Particulate emissions from indirect and secondary sources are as follows:

- Sulfur compounds
- Nitrogen compounds
- Hydrocarbon compounds

Even if the mechanism for their generation is understood in principle, indirect airborne particulate masses (generated from gases) can be quantified only with a high rate of uncertainty because the share of natural and anthropogenic precursors is rather uncertain.

Particulate emissions from direct natural sources are as follows:

- Volcanic activities
- Ocean salt
- Wind erosion
- Burning of biomass (caused by natural events such as lightning)

Particulate emissions from direct anthropogenic sources are as follows:

- Burning of biomass
- Waste incineration
- Stationary incineration
- Traffic (excluding road dust)
- Dust from construction, quarries, and mines
- Dust from agriculture

The size of particles from natural and anthropogenic sources can range from several nanometers (nm) up to several millimeters (mm). Figure 37 shows examples of particles and their sizes.

Large dust particles with diameters of more than 10 μm will quickly be deposited close to their sources. However, fine particles remain airborne for a certain time. Particles of less than 0.1 to 0.2 μm act as condensation nuclei in the formation of clouds and can be removed from the atmosphere only by rain. If such fine particles escape from the lower troposphere to higher air regions, they may remain there for months or even years [3].

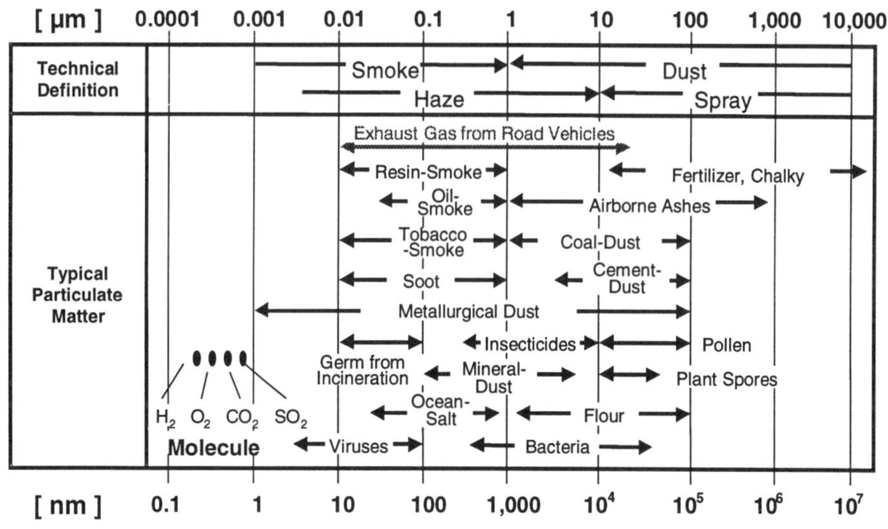

Figure 37. Size ranges of different types of particulate matter [84–86].

4.4.1 Airborne Dust Emissions in Europe

Available literature did not allow the establishment of a balance for total airborne dust emissions in Europe. However, one reference [134] estimated PM_{10} emissions for the European Union (Figure 38). Because combustion processes mainly generate small particles (Figure 37), the emission balance shows relatively high shares for these emission sources. Handling of bulk

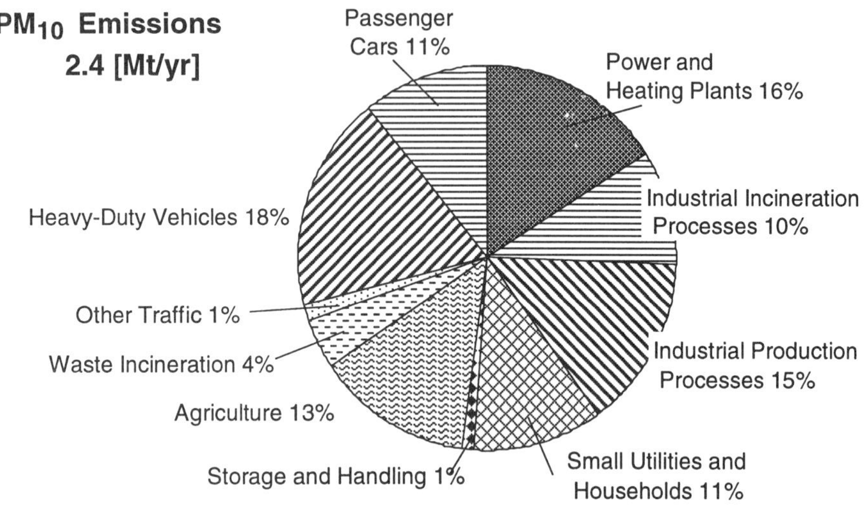

Figure 38. Contribution of various sources to PM_{10} emissions in the European Union (reference year 1993) [134].

goods is primarily responsible for the emission of particles greater than 10 μm and therefore is negligible for the PM_{10} emission balance. Emissions from cattle growing are connected with high uncertainties and may have been over-estimated here.

4.4.2 Airborne Dust Emissions in the Federal Republic of Germany

The decline in airborne dust emissions can be related mainly to the closing of outdated incineration facilities and industrial plants in the "new" states of the Federal Republic of Germany (Figure 39).

In addition, a short-term improvement of the efficiency of existing filtering units in power and heating plants was achieved. Conversion from solid to low-emission liquid and gaseous fuels, especially in small incineration units, also had a positive effect on the development of airborne dust emissions.

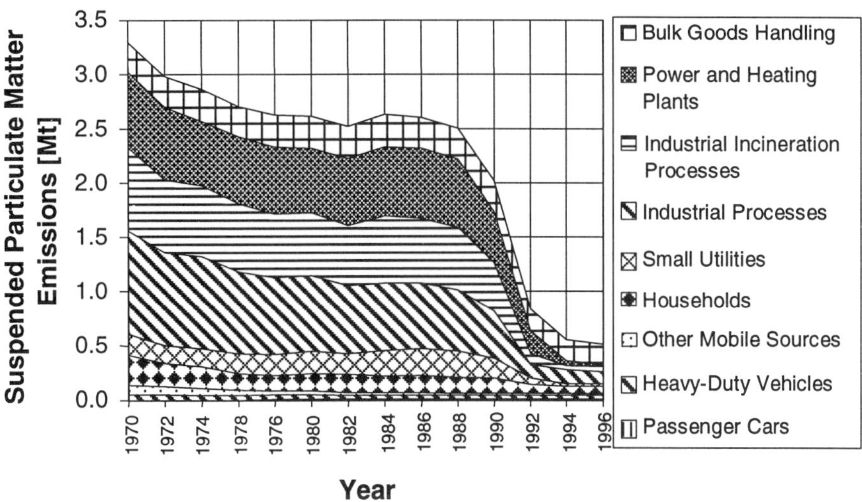

Figure 39. Trend for suspended particulate matter (SPM) emissions in the Federal Republic of Germany. (The contribution of passenger car traffic is so small that it cannot be shown within the scale of this graph.) [69, personal calculations]

The contribution of road traffic to these emissions is a marginal 8%. Industrial production processes and the handling of bulk goods are the primary emitters, together contributing almost 60% (Figure 40).

4.4.3 Judgment Criteria for the Impact of Airborne Dust on Air Quality

The EU-Directive 80/779/EEC determines in its Appendix IV two limit values which are specified as such in the "22nd Federal Air Quality Protection Ordinance" (BImSchV). This Ordinance transposes the EU-Directive into national law. A difference in the limits mentioned in the "Technical Guideline—Air" exists only insofar as the sum frequency value of 300 $\mu g/m^3$ is specified as 98% value in the Guideline and as a 95% value in the Directive (Table 7). In addition, the World Health Organization (WHO) has established a guide value of 120 $\mu g/m^3$ for the protection of human health.

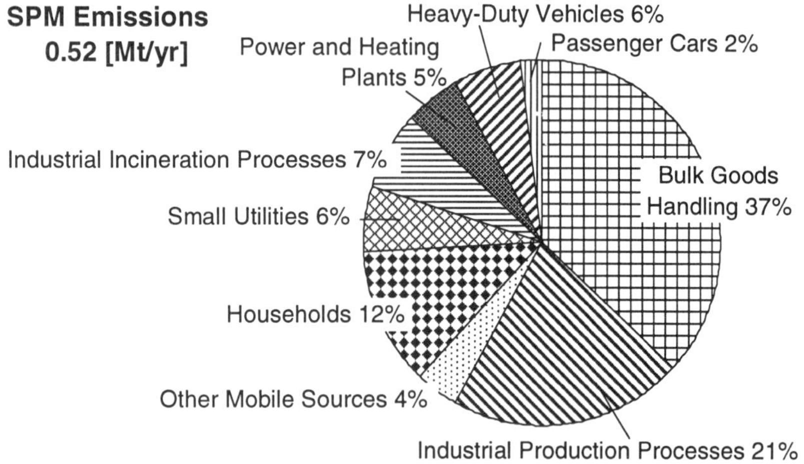

Figure 40. Contribution of various sources to suspended particulate matter (SPM) emissions in the Federal Republic of Germany (reference year 1996) [69, personal calculations].

TABLE 7. GUIDE AND LIMIT VALUES FOR SUSPENDED PARTICULATE MATTER (SPM) CONCENTRATIONS IN THE ATMOSPHERE IN THE FEDERAL REPUBLIC OF GERMANY

Suspended Particulate Matter			
EU-Directive 80/779/EEC, Appendix IV			
Concentration Value	**Reference Time Frame**	**Protection Objective**	**Character of Value**
150 μg/m^3	Arithmetic Yearly Average from Daily Average Values	Human Health	Limit Value
135 μg/m^3	95%-Value of Sum Frequency of Daily Average Values for One Year	Human Health	Limit Value
22nd Federal Air Quality Protection Law			
Concentration Value	**Reference Time Frame**	**Protection Objective**	**Character of Value**
150 μg/m^3	Arithmetic Yearly Average from Daily Average Values	Human Health	Limit Value
300 μg/m^3	95%-Value of Sum Frequency of Daily Average Values for One Year	Human Health	Limit Value

TABLE 7. (cont.)

Technical Guideline—Air			
Concentration Value	**Reference Time Frame**	**Protection Objective**	**Character of Value**
150 μg/m³	Arithmetic Yearly Average from Daily Average Values	Human Health	Limit Value
300 μg/m³	98%-Value of Sum Frequency of Daily Average Values for One Year	Human Health	Limit Value
VDI-Guideline 2310 (MIK-Value)			
Concentration Value	**Reference Time Frame**	**Protection Objective**	**Character of Value**
500 μg/m³	One-Hour Average Value of up to Three Consecutive Hours	Human Health	Guide Value
250 μg/m³	Daily Average Value	Human Health	Guide Value
150 μg/m³	Daily Average of Consecutive Days	Human Health	Guide Value
75 μg/m³	Yearly Average	Human Health	Guide Value
WHO			
Concentration Value	**Reference Time Frame**	**Protection Objective**	**Character of Value**
120 μg/m³	Daily Average	Human Health	Guide Value
Soot			
23rd Federal Air Quality Protection Law			
Concentration Value	**Reference Time Frame**	**Protection Objective**	**Character of Value**
8 μg/m³	Arithmetic Yearly Average	Human Health	Guide Value

Separate limit values for soot concentrations have been established in the course of the 23rd Ordinance to §40, Sec. 2 of the Federal Air Quality Protection Law. The limit value that became effective as of July 1, 1998 is shown in Table 7.

According to §40, Sec. 2 of the Federal Air Quality Protection Law, the Road Traffic Authority may reduce or even stop road traffic on certain roads or in certain areas as defined in the relevant legal provisions, in order to reduce detrimental environmental impacts or to prevent their occurrence [138]. Due to the development of soot concentrations in the atmosphere (see Section 4.4.4 of this chapter) and benzene (see Section 4.7.3 of this chapter), it is anticipated that traffic-reducing measures may be announced to meet applicable limit values at direct roadside locations.

4.4.4 Concentrations of Airborne Dust in the Atmosphere

The emission reduction achieved through control measures is reflected in the results of measurements of atmospheric concentrations, as shown in Figure 41. Even at roadside stations, measured values remain below the limit value of the "Technical Guideline—Air" by 50%. The concentrations at urban background measuring stations are approximately twice as high as the background concentrations caused by natural sources such as erosion (Figure 42) [69].

With regard to road traffic, particulate emissions from passenger cars and heavy-duty vehicles with diesel engines mainly are discussed here. These emissions usually are mistakenly referred to as soot emissions, although soot is only one component of particulates.

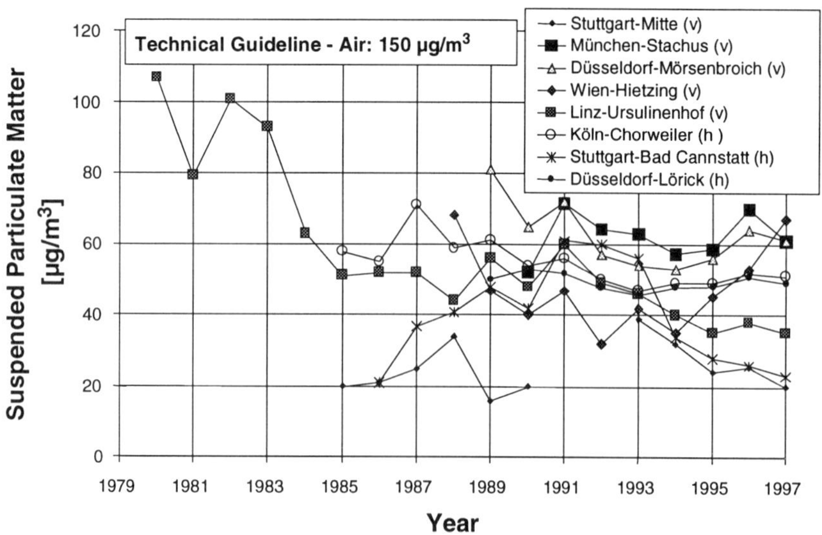

Figure 41. Development of suspended particulate matter (SPM) concentrations in the atmosphere at roadside measuring stations (v) and urban background measuring stations (h) in the Federal Republic of Germany and Austria [78–80, 83, 89, 131].

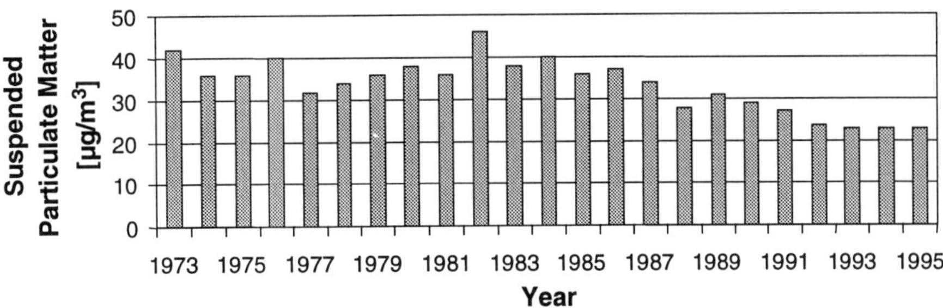

Figure 42. Development of suspended particulate matter (SPM) concentrations in the atmosphere at rural measuring stations of the Federal Ministry of Germany [69].

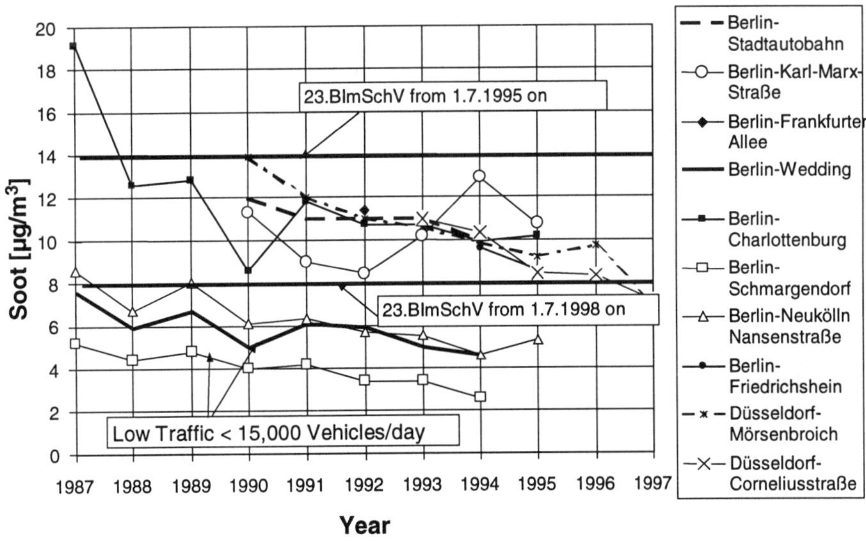

Figure 43. Development of soot concentrations in the atmosphere at roadside measuring stations in the Federal Republic of Germany [80, 131].

In connection with concentration levels in the atmosphere, soot is understood as the elementary carbon part of airborne particles. Figure 43 shows the development of soot concentrations in the atmosphere in two German cities, together with the limit values as specified in the 23rd Ordinance of the Federal Air Quality Protection Law.

The yearly average values show a decreasing trend at all measuring stations, while concentration values exceed the presently valid limit value only slightly at some high-traffic locations.

4.5 Carbon Monoxide

Carbon monoxide (CO) is generated during the incomplete combustion of carbon-containing fuels (i.e., processes lacking a sufficient supply of oxygen) and during the decay of organic material.

Carbon monoxide gas is colorless, odorless, and tasteless. According to one reference [8], carbon monoxide has an average lifetime of two months, a time that is too short for the generation of a globally uniform CO concentration. The CO concentration ranges from 100 to 200 ppb in the northern hemisphere, and from 50 to 80 ppb in the southern hemisphere. The depletion of CO occurs through OH radicals, a process in which CO and CH_4 "compete" with each other insofar as increasing CO emissions may hamper the depletion of methane [22]. Depletion of carbon monoxide also may occur during microbiological processes in the upper layers of the soil.

The low atmospheric CO concentrations today do not have negative effects on humans or vegetation.

4.5.1 Carbon Monoxide Emissions in Europe

The greatest uncertainties about the contribution to total CO emissions exist with regard to small industries and households. This can be explained in part by the effect of burning wood. Emission factors for small incineration units, especially when the units are operated by the burning of wood, vary by two orders of magnitude, and even the figures regarding the amount of firewood consumed differ largely. Figure 44 shows the contribution of various sources to total CO emissions.

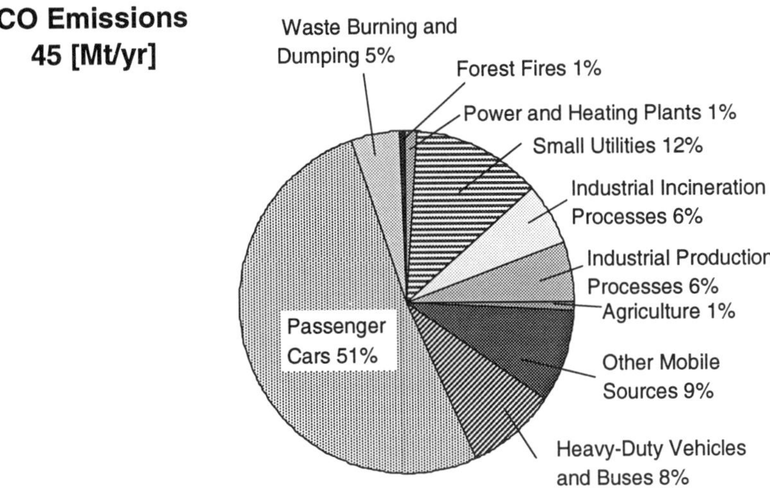

Figure 44. Contribution of various sources to carbon monoxide (CO) emissions in the European Union—15 States (reference year 1994) [71].

4.5.2 Carbon Monoxide Emissions in the Federal Republic of Germany

Carbon monoxide emissions have decreased since 1972, as shown in Figure 45, primarily because of legal requirements for the reduction of CO emissions from road traffic and the conversion of small incineration units to liquid and gaseous fuels.

Relatively considered, road traffic remains the major source of CO emissions in the Federal Republic of Germany (Figure 46). However, the contribution of this sector will continue to decrease because of increasingly stringent emission standards and the continued substitution of cars without catalysts by vehicles with three-way catalysts and low-emission diesel vehicles. Today, passenger cars classified as low-emission vehicles already have reached a share of 78% of the total passenger car fleet in the Federal Republic of Germany.

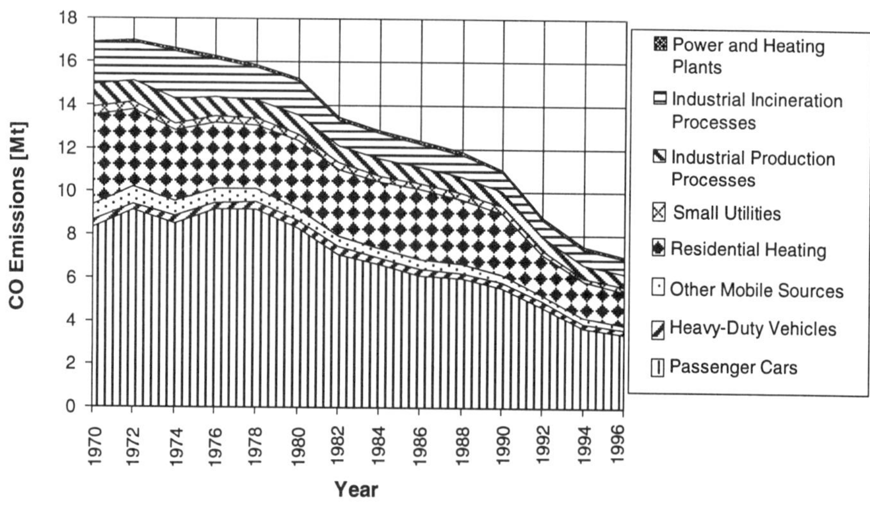

Figure 45. Trend of carbon monoxide (CO) emissions in the Federal Republic of Germany [69, personal calculations].

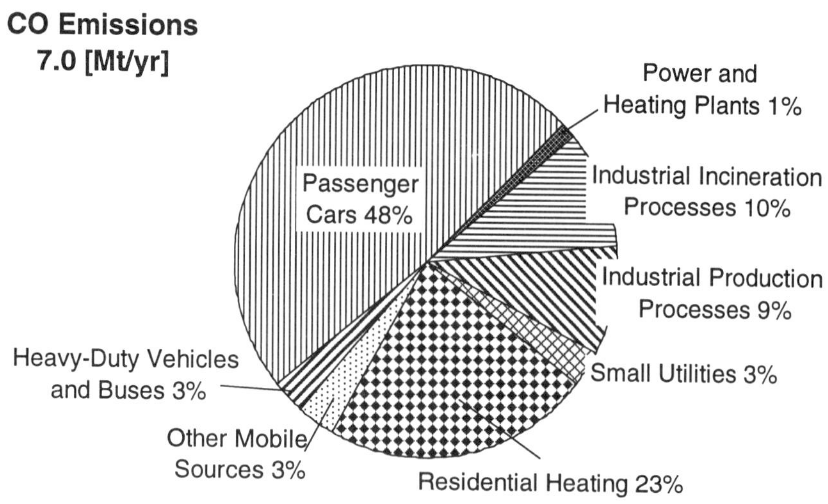

Figure 46. Contribution of various sources to carbon monoxide (CO) emissions in the Federal Republic of Germany (reference year 1996) [69, personal calculations].

4.5.3 Judgment Criteria for the Impact of Carbon Monoxide on Air Quality

Limit values for CO concentrations in the atmosphere are specified in the "Technical Guideline—Air." Guide values are given by the Commission for the Purification of the Air, of the Association of German Engineers (VDI-Kommission Reinhaltung der Luft) and by the World Health Organization (WHO); however, there are no relevant EU-Directives (Table 8).

TABLE 8. LIMIT AND GUIDE VALUES FOR CARBON MONOXIDE (CO) CONCENTRATIONS IN THE ATMOSPHERE IN THE FEDERAL REPUBLIC OF GERMANY

Carbon Monoxide			
Technical Guideline—Air			
Concentration Value	**Reference Time Frame**	**Protection Objective**	**Character of Value**
10 mg/m^3	Arithmetic Yearly Average of Daily Average Values	Human Health	Limit Value
30 mg/m^3	98%-Value of Sum Frequency of Daily Averages in One Year	Human Health	Limit Value
VDI-Guideline 2310 (MIK-Value)			
Concentration Value	**Reference Time Frame**	**Protection Objective**	**Character of Value**
50 mg/m^3	One-Hour Average Value during up to Three Consecutive Hours	Human Health	Guide Value
10 mg/m^3	Daily Average	Human Health	Guide Value
10 mg/m^3	Yearly Average	Human Health	Guide Value
WHO			
Concentration Value	**Reference Time Frame**	**Protection Objective**	**Character of Value**
60 mg/m^3	Half-Hour Average Value	Human Health	Guide Value
30 mg/m^3	One-Hour Average	Human Health	Guide Value
10 mg/m^3	Eight-Hour Average	Human Health	Guide Value

4.5.4 Concentrations of Carbon Monoxide in the Atmosphere

Measurements from roadside stations clearly show a decreasing trend for atmospheric CO. Figure 47 shows a decrease in CO concentrations at the

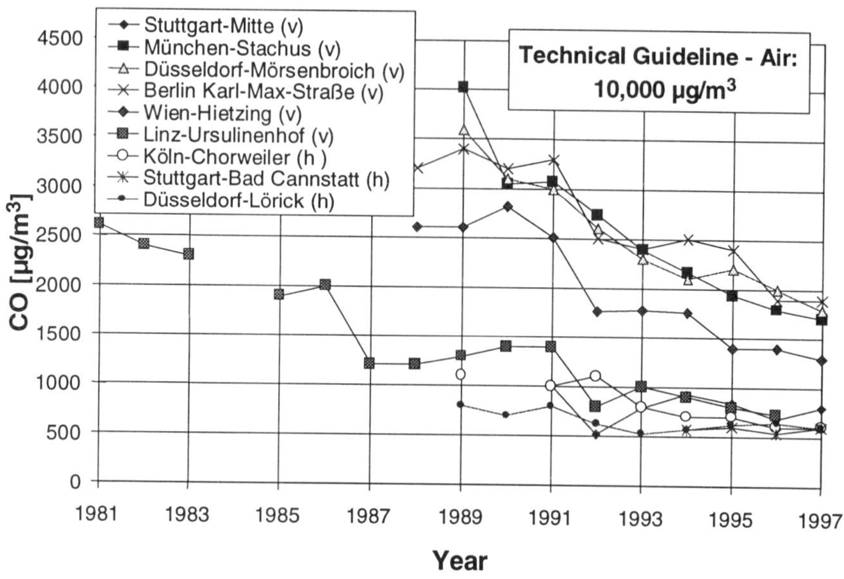

Figure 47. Development of carbon monoxide (CO) concentrations in the atmosphere at roadside measuring stations (v) and urban background measuring stations (h) in the Federal Republic of Germany and Austria [78–80, 83, 89, 131].

roadside measuring stations of almost 50%. The development of urban background concentrations demonstrates the progress made in home heating technologies.

4.6 Ozone

The ozone layer in the upper region of the atmosphere, referred to as the stratosphere, protects life on earth against the UV-B portion of sunlight. In the lower region of the atmosphere, referred to as the troposphere, ozone in excessive concentrations is an undesirable agent because it can irritate the respiratory tract and damage plants by penetrating the leaves.

Ozone (O_3) does not play a role in automobile exhaust gases. Only air-chemical processes triggered by radiation from the sun can cause the formation of ozone precursors and finally of ozone itself. The phenomenon, generally called "summer smog," is caused by a complicated interrelation of many chemical and meteorological components. This ozone-forming chemistry has not been fully explained to date.

Ozone precursors are volatile organic gases such as hydrocarbons, nitrogen oxides (NO_x), and carbon monoxide (CO), which can reach high concentrations, especially in congested areas. Via highly reactive intermediate products such as nitric oxide (NO), hydroxyl-radicals (HO), hydroperoxi-radicals (HO_2), and alkylperoxi-radicals (RO_2), ozone eventually will form.

4.6.1 Judgment Criteria for the Impact of Ozone on Air Quality

The EU-Directive 92/72/EEC determines in its Appendix IV two limit values which are specified as such in the "22nd Federal Air Quality Protection Ordinance" (BImSchV). This Ordinance transposes the EU-Directive into national law (Table 9).

Traffic restrictions in areas having increased ozone concentrations became possible as of July 26, 1995, through an amendment ("ozone law") to the Federal Air Quality Protection Law. According to this law, the operation of gasoline engine vehicles without catalysts and non-low-emission diesel vehicles may be prohibited (unless a special operation permit is granted) if ambient ozone concentrations reach an average hourly value of 240 $\mu g/m^3$ at least at three measuring stations in the Federal Republic of Germany, located more than 50 km (31 miles) but less than 250 km (155 miles) from each other, and if the ozone concentrations can be expected to reach the same ozone levels on the following day.

TABLE 9. GUIDE AND LIMIT VALUES FOR OZONE (O_3) CONCENTRATIONS IN THE ATMOSPHERE IN THE FEDERAL REPUBLIC OF GERMANY

Ozone			
EU-Directive 92/72EEC			
Concentration Value	**Reference Time Frame**	**Protection Objective**	**Character of Value**
200 µg/m³	98%-Value of Sum Frequency of Half-Hour Average Value for One Year	Human Health	Limit Value
110 µg/m³	Eight-Hour Average	Human Health	Limit Value
200 µg/m³	One-Hour Average	Vegetation	Limit Value
65 µg/m³	Daily Average	Vegetation	Limit Value
180 µg/m³	One-Hour Average	Human Health	Limit Value (Information for the Population)
360 µg/m³	One-Hour Average	Human Health	Limit Value (Warning of the Population)
Federal Air Quality Protection Law "Ozone Law"			
Concentration Value	**Reference Time Frame**	**Protection Objective**	**Character of Value**
240 µg/m³	One-Hour Average	Human Health	Limit Value
VDI-Guideline 2310			
Concentration Value	**Reference Time Frame**	**Protection Objective**	**Character of Value**
120 µg/m³	Half-Hour Average	Human Health	Guide Value
WHO			
Concentration Value	**Reference Time Frame**	**Protection Objective**	**Character of Value**
120 µg/m³	Eight-Hour Average	Human Health	Guide Value

4.6.2 Concentrations of Ozone in the Atmosphere

Figures 48 and 49 show the development of the yearly average values for ozone concentrations in the atmosphere.

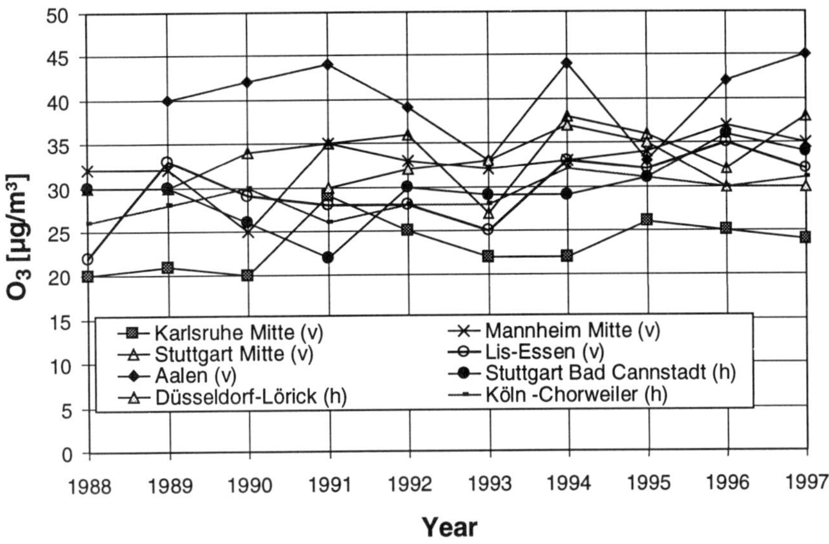

Figure 48. Trend of ozone (O_3) concentrations in the atmosphere at roadside measuring stations (v) and urban background measuring stations (h) in the Federal Republic of Germany [80, 89].

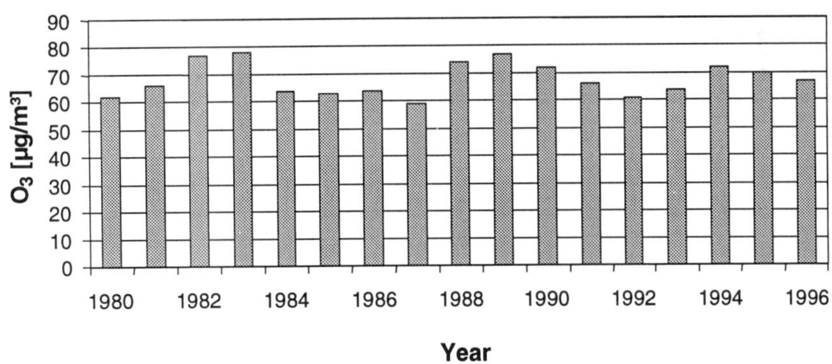

Figure 49. Trend of ozone (O_3) concentrations in the atmosphere at rural measuring stations of the Federal Ministry of Germany [69].

A uniform trend cannot be detected because the yearly averages are affected heavily by varying weather conditions such as hot, cool, or rainy summers, or by changing wind directions. Large-scale transport mechanisms in horizontal and vertical directions further contribute to widely differing measurement results. Weather conditions also are the most influential factors at rural measuring stations.

Conditions of maximum ozone impact may be derived from the maximum one-hour value, which has decreased from 336 $\mu g/m^3$ in 1990 to 257 $\mu g/m^3$ in 1997, or by the number of days on which the threshold value for warning the public is being exceeded [69].

4.7 Benzene

Benzene (C_6H_6) belongs to the group of non-methane hydrocarbons. Due to the environmental and political significance that this substance had gained through the "23rd Federal Air Quality Protection Ordinance," benzene will be considered here in a separate section.

Benzene, the simplest aromatic hydrocarbon, is a colorless liquid with a characteristic odor. It is barely soluble in water but can be mixed with most organic solvents. Benzene is used as a solvent and extraction media, as well as for the production of numerous substances that are base products for synthetic material, medications, and colors. Benzene is stable at room temperature. If it exists in vapor form in the atmosphere, it will be depleted completely through natural processes [88].

Benzene depletion in the atmosphere occurs primarily by means of a process known as indirect phototransformation. During this process, benzene reacts with a variety of photochemically formed reactive molecules or molecule parts. The most important of these reactions is the transformation of OH radicals, which occurs at a relatively slow speed. By comparison, the reaction is five times slower than with toluene and 19 times slower than with m-xylene.

No publications were available from which a benzene emission balance could have been determined for the European Union.

4.7.1 Benzene Emissions in the Federal Republic of Germany

The largest benzene emissions in industrialized nations come from road traffic. Figure 50 shows the contribution of various sources to total benzene emissions in the Federal Republic of Germany for the reference year 1995.

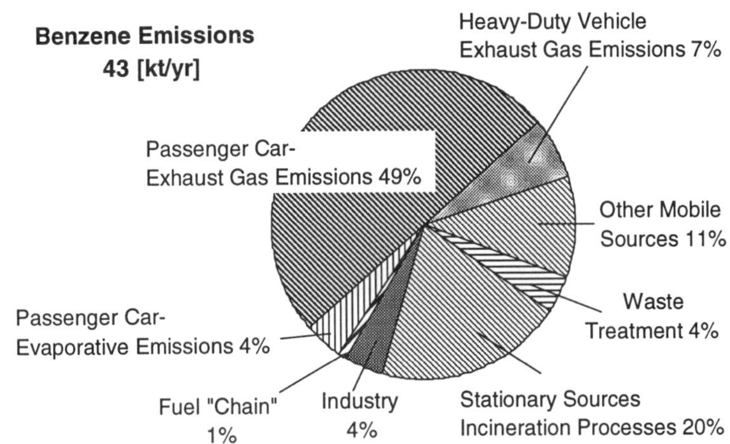

Figure 50. Contribution of various sources to anthropogenic benzene emissions in the Federal Republic of Germany (reference year 1995) [47, 90, personal calculations].

The values mentioned for waste utilization, stationary incineration, and industry are connected with high uncertainties and may be substantially higher than mentioned here. Waste utilization includes gaseous emissions from dumps and benzene emissions from waste incineration. Emissions generated by incineration processes at households, small utilities, and large incineration units are grouped into the sector of stationary incineration. The sector industry contains emissions from coke ovens, the oil industry, and solvent usage.

4.7.2 Judgment Criteria for the Impact of Benzene on Air Quality

The 23rd Ordinance to §40, Sec. 2 of the Federal Air Quality Protection Law defines guide values for roadside concentrations (Table 10). To achieve the envisaged air quality target, the federal government may impose strict countermeasures such as traffic restrictions.

The State Commission for the Protection of the Air (**L**änder-**A**usschuß für **I**mmissionsschutz—LAI) is working on judgment criteria for determining the cancer risk attributable to air pollution. Because a reference value cannot be defined for all substances that supposedly cause cancer, the LAI selected seven substituting components, one of which is benzene.

TABLE 10. LIMIT AND GUIDE VALUES FOR BENZENE CONCENTRATIONS IN THE ATMOSPHERE IN THE FEDERAL REPUBLIC OF GERMANY

Benzene			
23rd Federal Air Quality Protection Regulation			
Concentration Value	**Reference Time Frame**	**Protection Objective**	**Character of Value**
15 μg/m³	Yearly Average (as of July 1, 1995)	Human Health	Guide Value
10 μg/m³	Yearly Average (as of July 1, 1998)	Human Health	Guide Value
State Committee for Air Quality (LAI)			
Concentration Value	**Reference Time Frame**	**Protection Objective**	**Character of Value**
2.5 μg/m³	Yearly Average	Human Health	Assessment Value

According to calculations made by the LAI, the cancer risk for a hypothetical population subjected to the combined effect of the concentration levels mentioned in Table 11 over 70 years is 1:2,500 [94]. In this regard, note that the cancer risk from air pollution is only 2%. Approximately 98% of the cancer risk is attributed to other factors such as genetic disposition and living habits, with the negative effects of cigarette smoking being in first place [80].

TABLE 11. JUDGMENT CRITERIA FOR THE SEVEN SUBSTANCES SELECTED BY THE LAI TO REPRESENT SUSPECTED CANCER-CAUSING COMPONENTS [94]

Component	Judgment Criteria for a Total Risk of 1:2,500*
Arsenic and Its Organic Compounds	5 ng/m^3
Asbestos Fibers	88 fibers/m^3
Benzene	2.5 μg/m^3
Cadmium and Its Compounds	1.7 μg/m^3
Diesel Engine Emissions	1.5 μg/m^3
PAH with Guide Substance Benzo[a]pyrene	1.3 μg/m^3
Dioxin with Guide Substance 2,3,7,8-TCDD	16 fg/m^3

4.7.3 Concentrations of Benzene in the Atmosphere

Because of substantial emission reductions achieved through the ongoing conversion of the existing car population to vehicles with three-way catalyst technology, atmospheric roadside benzene concentrations have shown a decreasing trend in recent years. As an example, Figure 51 shows the development of atmospheric benzene concentrations expressed as arithmetic yearly averages at three roadside measuring stations and one urban background measuring station in the Federal Republic of Germany. The atmospheric concentration at the urban background station remains more than 50% below the presently valid standard. At roadside measuring stations, it can be expected that the concentrations will remain below the limit value because of improved engine and emission control technologies.

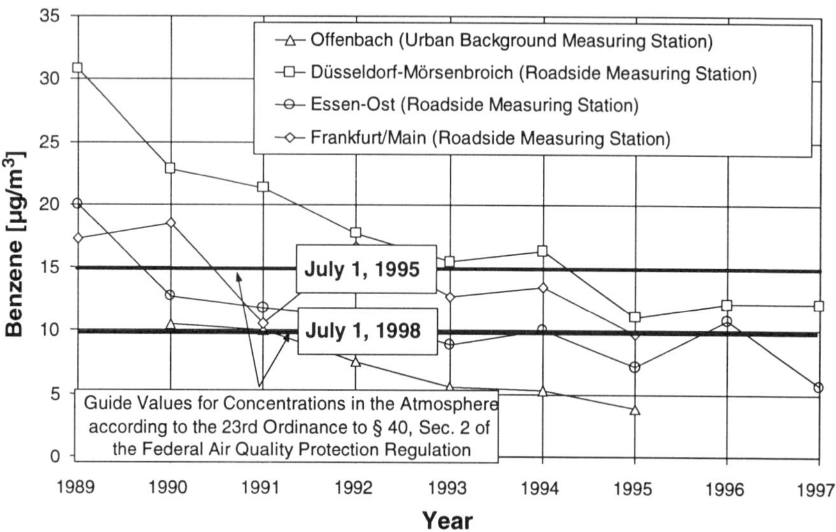

Figure 51. Trend for benzene concentrations in the atmosphere as arithmetic average from three roadside measuring stations and one urban background measuring station [80, 95].

4.8 Polycyclic Aromatic Hydrocarbons

Polycyclic aromatic hydrocarbons (PAHs) consist of condensed benzene rings and may also incorporate five-ring elements [96]. Their molecules contain only carbon and hydrogen atoms.

Polycyclic aromatic hydrocarbons are generated during incomplete burning of organic material and are emitted into the atmosphere in gaseous form or adsorbed on aerosols or solid particles. The PAH-profile (i.e., the mass ratio of individual PAHs to each other within a given mixture) may differ substantially, depending on the fuel used and the combustion process applied. These different PAH profiles mix in the ambient air at least in the long run and eventually will form similar profiles. For making a cancer risk assessment, the ideal case is assumed, namely, that the PAH profiles are identical and that it is acceptable to use the concentration of a single PAH as a measure of the carcinogenic potential of the full mixture [94]. For this assessment, benzo[a]pyrene is used as a guide component because this substance has been investigated the most and has the highest carcinogenic potential [94]. However, benzo[a]pyrene (BaP) also represents the most unstable PAH compound [87, 96], which makes its role as a guide substance questionable. Capillary gas chromatographic measurements have shown that atmospheric airborne dust contains many more than 100 different PAH compounds of largely differing carcinogenic potential. To obtain a correct judgment of the effects, it is necessary to clarify how much importance must be attributed to the other PAHs of the mixture. In nature, PAHs may be decomposed in the atmosphere through photolysis or by biological depletion [96].

As products from pyrolysis from organic material, PAHs have been determined to be proven carcinogens in animal tests. The Federal Republic of Germany has defined a "technical guide concentration" of 2 $\mu g/m^3$; for certain areas, this value is set higher at 5 $\mu g/m^3$ [148].

4.8.1 Concentrations of Polycyclic Aromatic Hydrocarbons in the Atmosphere

Measurements of PAH concentrations exist only for PAH concentrations in airborne dust [80]. Measurements for these concentrations are more or less limited to the determination of benzo[a]pyrene, benzo[e]pyrene, dibenzo[a,h] anthracene, benzo[ghi]perylene, benzo[a]anthracene, and coronene concentrations in airborne dust. No data are available for PAH sum concentrations.

Polycyclic aromatic hydrocarbon concentrations are subjected to pronounced variation, depending on the time of the year. In winter, the measured values may reach a level that is nine times higher than the level in summer months. In this regard, note that in small rural towns the concentration levels in ambient air during winter months are similar to those found during summer months. This may result from an increased use of fossil fuels, especially in households, during winter [97].

Figure 52 shows the development of benzo[a]pyrene and coronene concentrations in the Rhein/Ruhr area [80]. The data given are area averages developed from the yearly average value of all existing measurement stations of the LIMES (Länder-Immissions – Meβ-und Ermittlungssystem) net established by the State Environment Agency of Nordrhein-Westfalen.

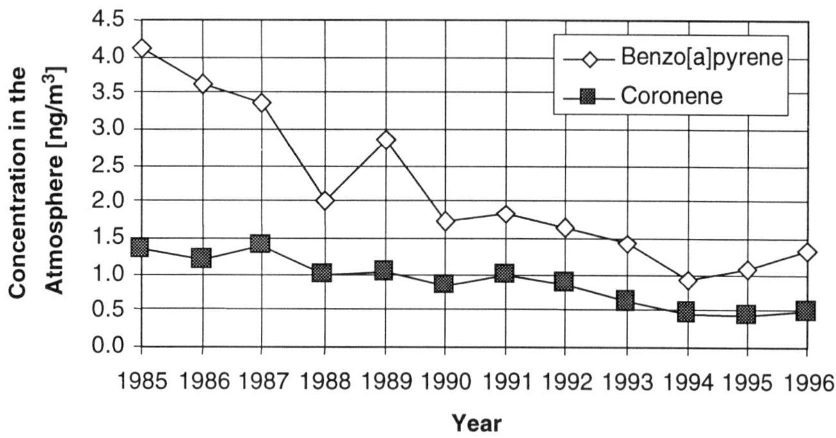

Figure 52. Development of benzo[a]pyrene (BaP) and coronene concentrations in the atmosphere as regional averages of arithmetic yearly mean values from the TEMES measuring grid [80].

CHAPTER 5

Possible Ways to Influence Emissions from Road Traffic—Emission Scenarios

Real field emissions from passenger car and heavy-duty vehicle traffic are determined not only by vehicle-specific conditions, which are partly prescribed by legislative requirements. They are also influenced to a substantial degree by the way the driver operates his or her car.

To allow a quantified judgment of both the effect of legislative measures (which include emission standards, fuel requirements, and in-use vehicle inspection) and driver influence, a computer program for calculating emissions from passenger car and heavy-duty vehicle traffic was developed at the Technical University of Vienna (Institute for Combustion Engines and Vehicle Design). This program and the results obtained by its application are presented in the following sections.

The difficulty with the calculation, which contains, in simple words, a determination of the vehicle populations and vehicle kilometers traveled and the combination of these data with the emission factors, is the correct selection and prognosis of the future development of these input parameters.

The different estimates for the expected growth of the vehicle population shown in Figure 53 demonstrate the difficulty in giving a reliable prognosis about the future development of road traffic. Actual development did exceed all projections.

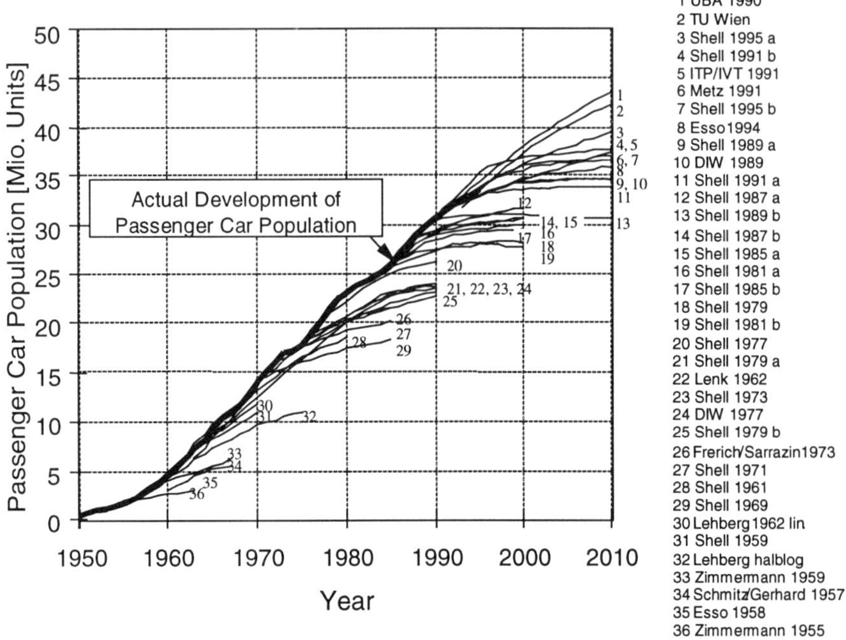

Figure 53. Comparison of estimates for the development of passenger car population from 1950 to 2010, with the actual car population development in the "old" states of the Federal Republic of Germany [98].

5.1 Calculation Model for Determining Emissions from Passenger Car and Heavy-Duty Vehicle Traffic in the Federal Republic of Germany

The computer program calculates passenger car and heavy-duty vehicle emissions in the "old" and "new" states of the Federal Republic of Germany from 1970 through 2020 for 44 passenger car categories. These categories are separated according to individual propulsion method, level of emission standards met, and individual displacement class, as well as for 142 heavy-duty vehicle categories separated according to individual propulsion method, level of emission standards met, and individual weight class. The determination of the future growth of the vehicle population is based on a trend calculation for the expected newly registered vehicles and on the "survival probability" of vehicles extrapolated from the year 1996 (Figures 54 and 55). Investigations from Shell [99], Esso [100], and the ifo-institute for economic studies [101] have been used for the trend calculations.

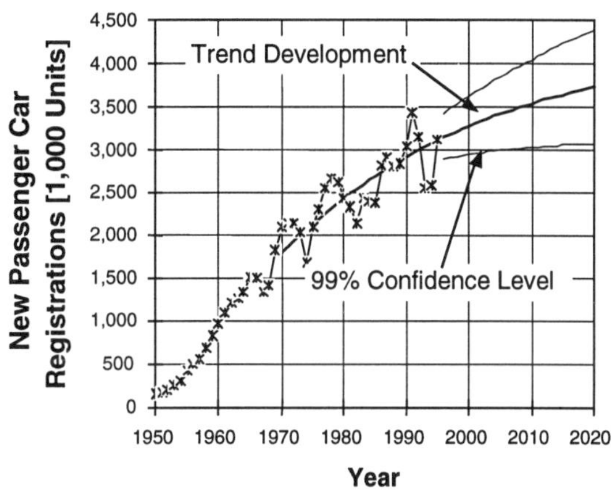

Figure 54. New passenger car registrations in the "old" states of the Federal Republic of Germany from 1950 to 2020.

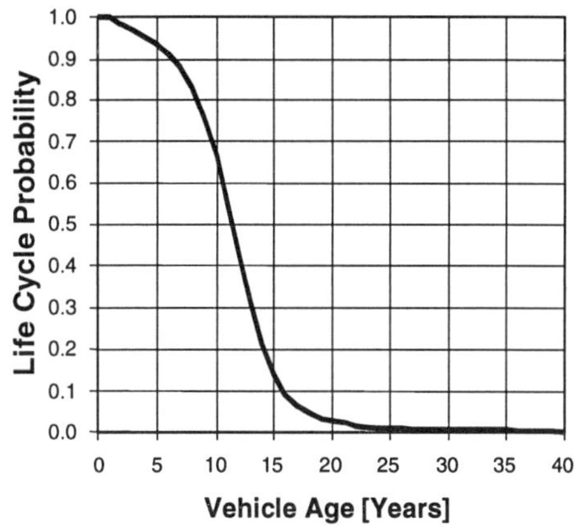

Figure 55. Passenger car life cycle probability as a function of vehicle age for the "old" states of the Federal Republic of Germany.

The separation of the car fleet into the different vehicle categories is being calculated from the individual new registrations and a certain life cycle probability. The resulting development of the car fleet is shown in Figures 56 and 57. In the interest of clarity, these graphs include neither the differentiation into displacement classes for passenger cars nor into weight classes and level of emission standards for heavy-duty vehicles.

Fuel consumption data and emission factors for nitrogen oxides, total hydrocarbons, carbon monoxide, and particulate matter were taken from a literature source [102] for the reference year 1990. These emission factors, provided by TÜV Rheinland (Technischer Überwachungs-Verein) for individual vehicle categories, were developed each for a vehicle or engine group with a certain average mileage.

With these data pairs—reference year and corresponding mileage—and by means of the relationship between exhaust emission and average vehicle mileage calculated from the information taken from data banks [81], it was possible to generate the specific emissions in relation to reference year and vehicle age. Furthermore, this data bank was used to determine emission factors for future emission control technologies.

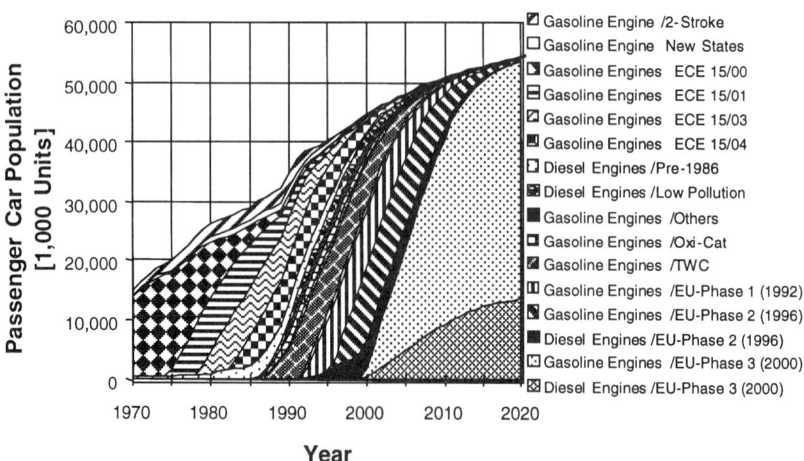

Figure 56. Former and future compositions of the vehicle population in the Federal Republic of Germany from 1979 to 2020.

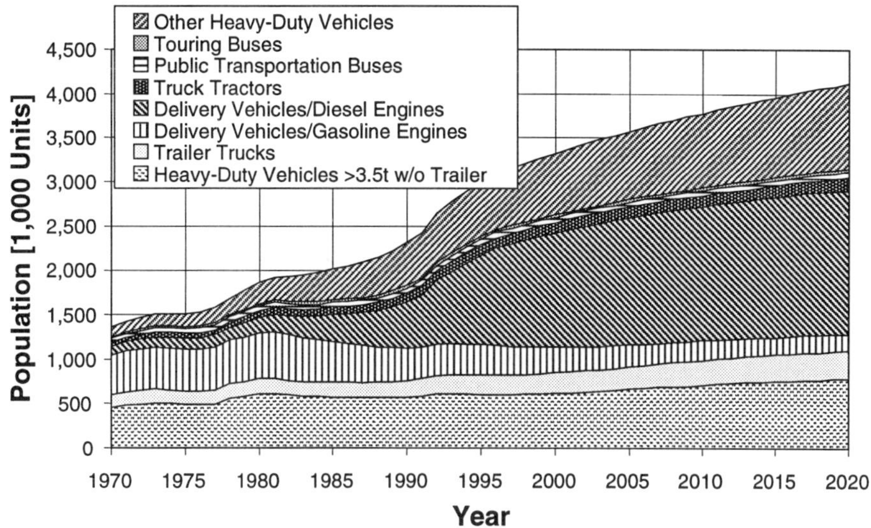

Figure 57. Former and future development of the composition of the heavy-duty vehicle population in the Federal Republic of Germany from 1970 to 2020.

The driving cycles representative for urban and extra-urban (rural) roads as well as for driving on federal highways (Bundes-Autobahn) were formed by taking into account the information contained in two literature sources [103, 104] and by using the driving sequences given in one literature source [102] in such a way that the average speeds of real traffic were simulated approximately. Verification of the validity of the established driving sequences with their relevant emission factors, of the level of the average yearly mileage, of the vehicle age-depending mileage distribution, and of the systematic of the cold-start effect calculation was performed. This was done by comparing the calculated fuel consumption and the total amount of gasoline and diesel fuel used in the Federal Republic of Germany in a given year.

The computer model generally uses yearly emissions as the smallest calculation unit, whereas cold-start emissions are calculated on a monthly basis. The latter is justified because vehicle mileage accumulation is not equally distributed over the year and because of the non-linear relationship between cold-start temperature and emissions (Figures 58 and 59).

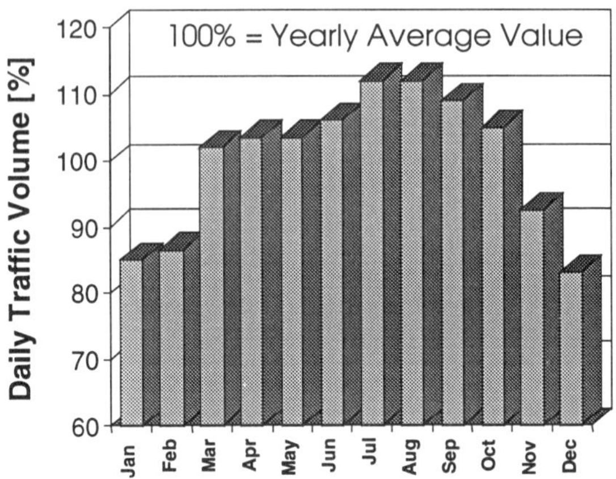

Figure 58. Average daily traffic volume (DTV) as mean value on federal highways A5, A8, and A81 (reference year 1991) [105].

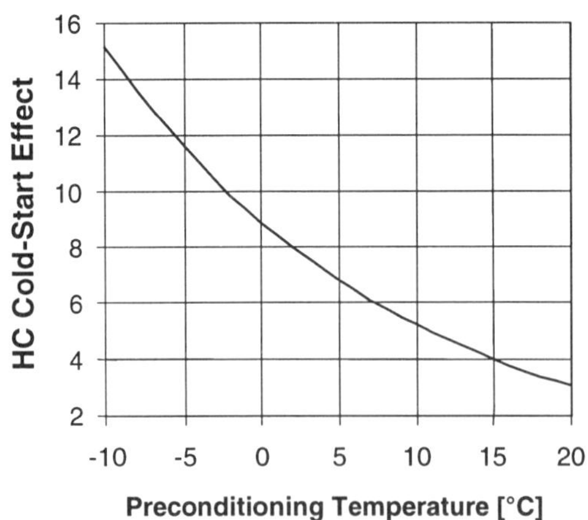

Figure 59. Cold-start effect for HC emissions from passenger cars with gasoline engines and three-way catalysts as a function of the preconditioning temperature [102].

The calculation method for determining the amount of emissions generated by a given passenger car category (e.g., vehicles with a gasoline engine, three-way catalyst, and a displacement of <1.4 liters) is shown in Figure 60. The method used for passenger cars is the same as the one used for heavy-duty vehicles, except for the calculation of cold-start emissions, which could not be incorporated into the computer model because of the lack of relevant investigations. The figure shows the calculation for cold-start and warm-start emissions.

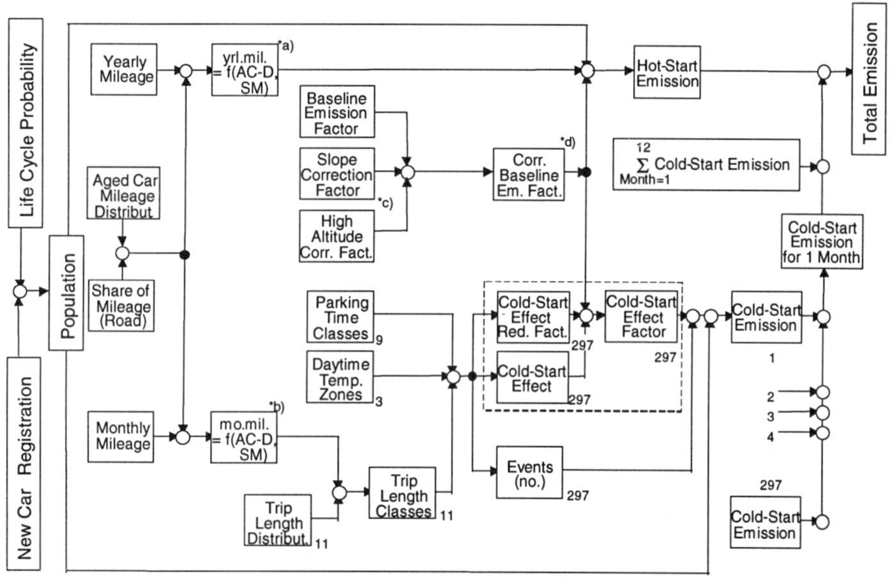

Figure 60. Calculation method for annual emissions from a designated vehicle category.

**a) yrl. mil.: Yearly mileage; AC-D: Aged car mileage distribution SM: Share of mileage on different road classes*

**b) mo. mil.: Monthly mileage; AC-D: aged car mileage distribution SM: Share of mileage on different road classes*

**c) Corr. Fact.: Correction factor*

**d) Em.-Fact.: Emission factor*

Based on new registration figures and "survival probabilities," the annual car population is calculated for each of the 44 passenger car categories and each of the 142 heavy-duty vehicle categories. The vehicle mileage figures are determined for the individual vehicle category and road type from the average mileage by means of the distribution of vehicle age and road type. The emissions generated by a given vehicle category on a given road type and for a given year under warmed-up engine conditions are calculated by multiplying the corresponding car population figure with the annual mileage and the emission factor corrected for slope and altitude.

Cold-start and warm-up emissions are considered only for the calculation of emissions generated on urban roads. The basis for the calculation of the monthly cold-start and warm-up emissions for each vehicle category is the classification of the annual average mileage with the trip length distribution into eleven trip-length classes. Other parameters of this calculation are the distribution of the daytime temperature and the differentiation into parking (soaking) time classes.

The hourly classification of the soaking time in intervals from zero to eight hours allows for individual consideration of cold-start and warm-up emissions, starting with a "real" cold start after eight or more soak hours and a hot start after zero to one hour of soaking time.

From this method of structuring into mileage classes, day temperature zones and soaking time classes result in 297 different cases. A case represents a start at a certain temperature, after a defined soaking time with a following trip over a specified distance (which is one of the eleven mileage classes).

The total monthly cold-start and warm-up emissions for a given vehicle category result from the multiplication of the "cold-start effect" with the number of events and the number of cars in the car population. Thereafter, all 297 cases are summed. Finally, cold-start and warm-up emissions of all 12 months are summed, and the result is added to the hot-start emissions of the given year, in order to arrive at the total annual emission of a given vehicle category.

5.2 Legislative Measures

Emission control regulations limit the extent to which automobiles impact air quality, especially by directly influencing the level of the emission factors. The emission characteristics of passenger cars in the field can be influenced by the following:

- Level of emission standards
- Required emission durability
- Fuel composition
- In-use inspection and maintenance

5.2.1 Level of Emission Standards

The impact of road traffic on air quality in California during the 1940s has already been discussed. The unique adverse atmospheric situation, especially in the Los Angeles basin, is known as California "summer smog." This situation also was aggravated by the exhaust and evaporative emissions from motor vehicles. Under "summer smog" conditions, chemical reactions of nitrogen oxides (NO_X) and hydrocarbons (HC) under intense influence of ultraviolet (UV) radiation and higher temperatures form photo-oxidants. In Europe, CO emissions from road traffic became a matter of concern in the 1960s because of the direct health risk for humans.

Legislative countermeasures started first in the United States, with the introduction of the 1961 emission control regulations in California. The first legislative auto emission control requirement in the European Community became effective in 1970. The further development of both regulatory frameworks is depicted in Figure 61 for the exhaust gas constituents carbon monoxide (CO) and the sum of hydrocarbons and nitrogen oxides (HC + NO_X).

Figure 62 shows the development of the corresponding emission standards for heavy-duty vehicles.

The effectiveness of these legislative requirements in reducing emissions from road traffic *vis-à-vis* the ever-increasing car population is evaluated in the following sections on the examples of nitrogen oxide, hydrocarbon, and carbon monoxide emissions from passenger cars.

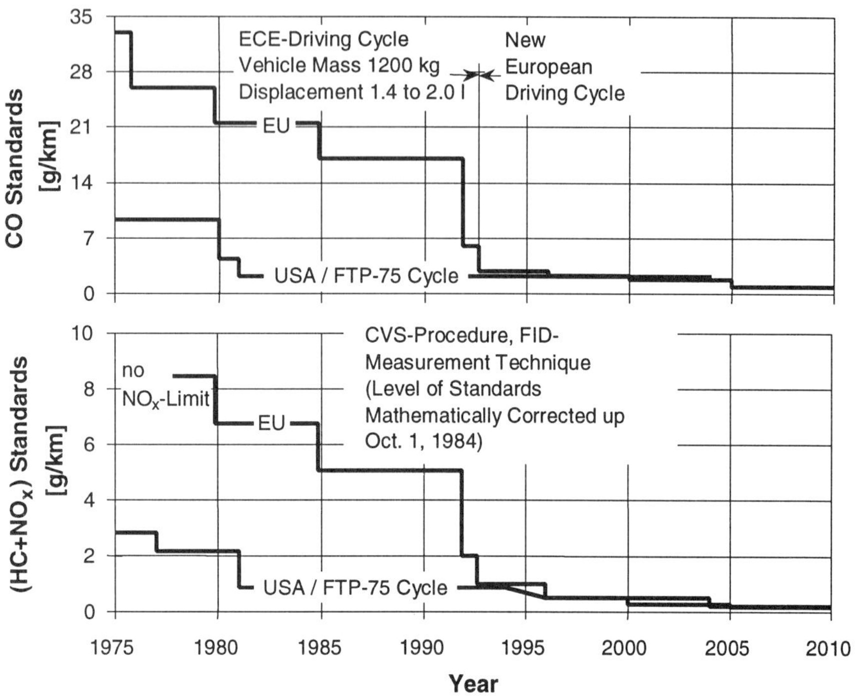

Figure 61. Evolution of exhaust emission standards for passenger cars with gasoline engines in the European Union (EU) and in the United States for carbon monoxide (CO) and the combined standard for hydrocarbons and nitrogen oxides (HC + NO_x) [81, 106].

Figures 63 through 65 show the emission development for nitrogen oxides, total hydrocarbons, and carbon monoxide under three different regulatory scenarios, together with the development of total vehicle miles traveled. Scenario "Euro 2" means that the emission factors are kept constant after the year 1996, or, in other words, emission control technology is frozen at the 1996 level. Under the same principle, scenario "Euro 3" is related to the year 2000 and scenario "Euro 4" to the year 2005, respectively.

As shown in Figures 63 through 65, all three exhaust emissions (NO_x, HC, and CO) will be reduced continuously beyond the year 2000 and eventually will be kept constant at a very low level, even under the assumption of further increasing vehicle miles traveled. When comparing the level achieved in the year 2020 with the starting point of the drop in the emission curves in 1987, the effectiveness of each legislative step becomes evident, as follows:

Legislation	**Improvement vs. 1987**		
	NO_x	**HC**	**CO**
"Euro 2"	78%	91%	76%
"Euro 3"	85%	95%	80%
"Euro 4"	91%	97%	85%

This comparison clearly shows the marginal additional air quality benefit achieved through the implementation of the "Euro 4" legislation, which, on the other hand, entails substantial technical efforts and increased costs for development of adequate emission control systems.

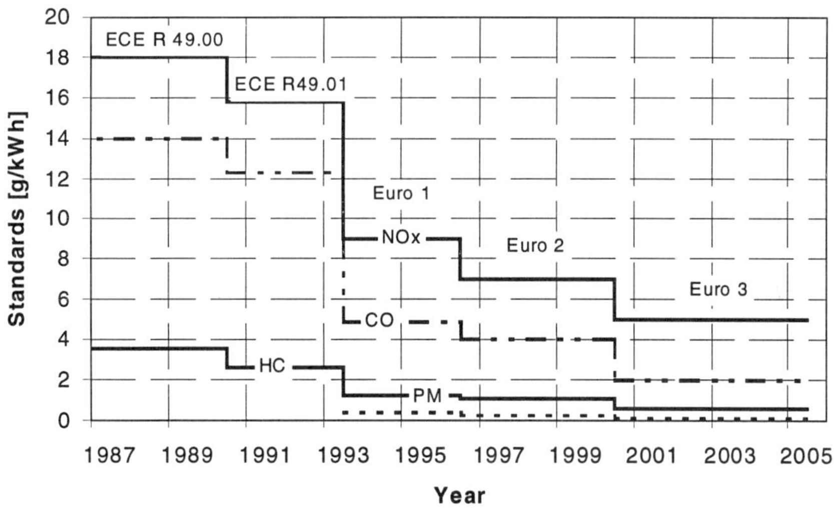

Figure 62. Evolution of assembly line exhaust emission standards for commercial vehicles with a vehicle mass of >3.5 tons and engine power of >85 kW in the European Union [108].

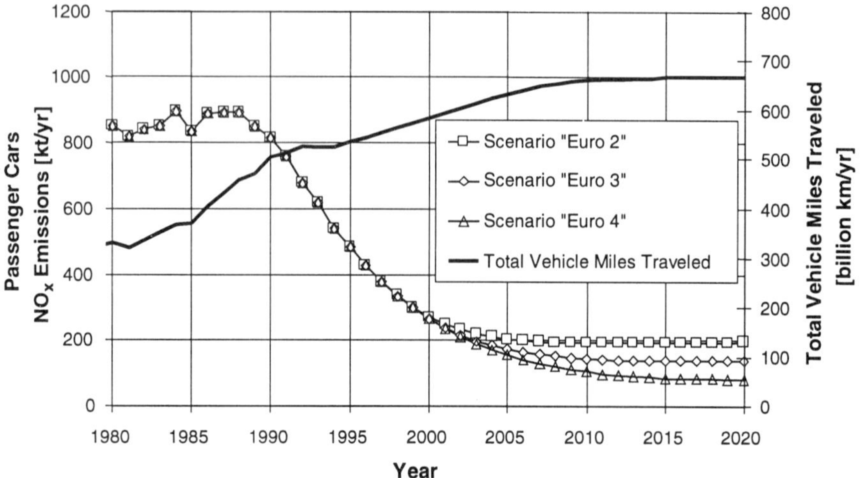

Figure 63. Development of passenger car nitrogen oxide (NO_x) emissions for various emission control regulations compared to the development of total vehicle miles traveled.

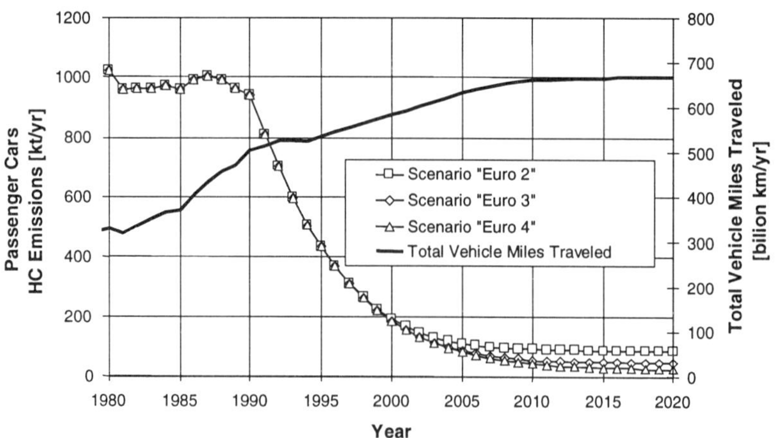

Figure 64. Development of passenger car hydrocarbon (HC) emissions for various emission control regulations compared to the development of total vehicle miles traveled.

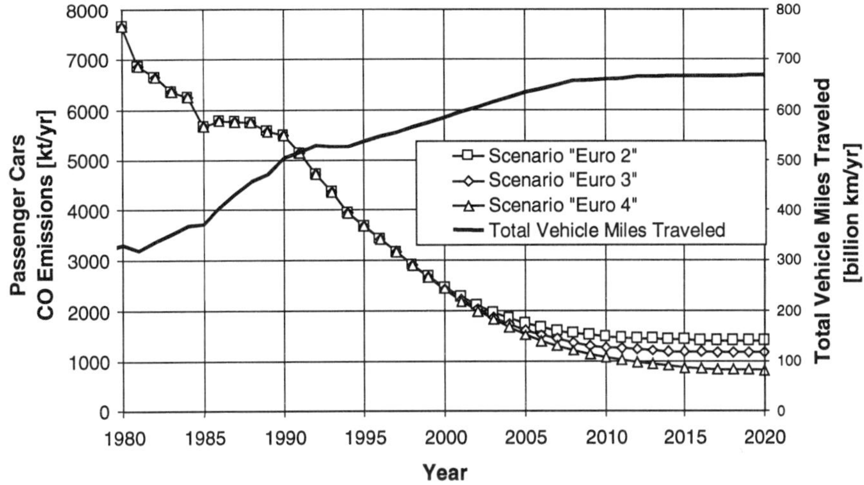

Figure 65. Development of passenger car carbon monoxide (CO) emissions for various emission control regulations compared to the development of total vehicle miles traveled.

The result of the calculations, which are shown graphically in the previous figures, can be summarized as follows:

- **"Euro 3" legislation ensures that a very low emission level will be achieved and maintained after full conversion of the existing car fleet to vehicles meeting the involved emission standards.**

- **"Euro 4" legislation achieves only a marginal further emission reduction compared to "Euro 3."**

5.2.2 In-Use Vehicle Emissions

The following section highlights an important but generally overlooked fact about the interrelationship between the application of increasingly stringent emission certification standards (or, in other words, the application of increasingly sophisticated and effective emission control systems) and the resulting emission durability in the field.

Without considering the individual production year (i.e., emission control technology applied), an emission deterioration of the vehicle mix in the field over mileage results is shown in Figure 66. The average deterioration is characterized by a (nonlinear) regression line with its 99% confidence interval, which describes the range in which the average HC emissions at a given mileage can be found with a probability of 99%.

However, if the production year (or, in other words, the applied emission control technology) was taken into account, we would see the effect as shown in Figure 67, namely, that together with the (expected) decreasing emission level, the deterioration factor for the emissions from vehicles of a given production year is being reduced. This means that

> **The introduction of increasingly stringent emission standards during the past years, which has triggered improved engine and emission control technologies, did not only lead to a lowering of the absolute emission level but also has entailed a more favorable long-term emission stability during the actual field operation of the vehicle.**

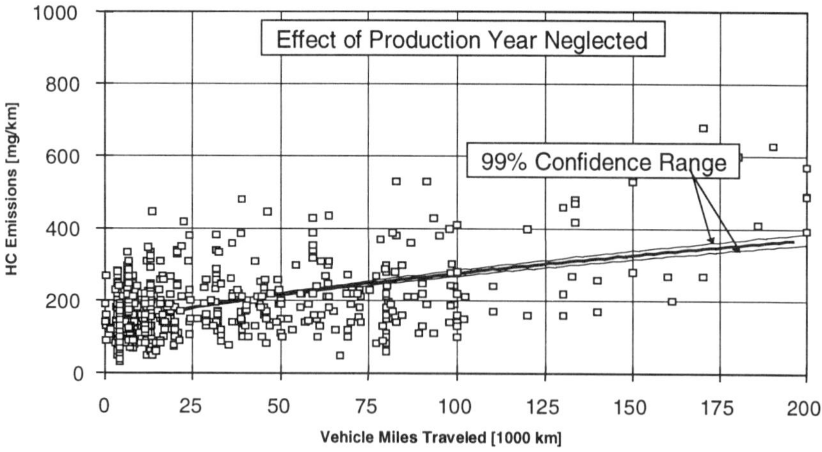

Figure 66. Deterioration of hydrocarbon (HC) emissions (FTP-75 results) from passenger cars with gasoline engines and three-way catalysts (evaluation of 3,671 measurements on 1,514 vehicles) [81].

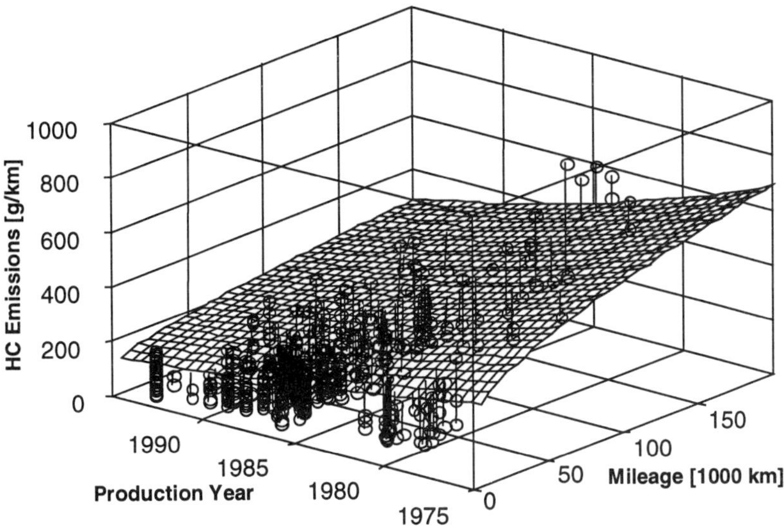

Figure 67. Effect of mileage and production year on hydrocarbon (HC) emissions (FTP-75 results) from passenger cars with gasoline engines and three-way catalysts (evaluation of 3,671 measurements on 1,514 vehicles) [81].

5.2.3 Fuel Composition

In Europe and the United States, legislative efforts toward improving air quality do not only lead to continuously increasing pressure on vehicle manufacturers to develop and introduce further improved emission control systems. They also trigger similarly demanding requirements for the oil industry toward producing advanced (i.e., low-emission) fuels.

Tables 12 and 13 show the present and future specifications established in the European Union for gasoline and diesel fuels [90, 109], together with the target values defined by the ACEA [107]. For comparison purposes, average values from market fuel analyses also are mentioned in these tables [93, 109].

The available emission reduction potential of different fuel compositions can be seen in Tables 14 and 15. An extensive literature review revealed that this emission reduction potential depends not only on the fuel composition but also on the applied engine and emission control technology. Furthermore, it is difficult to establish precisely the individual effect of a certain fuel specification because some specifications are interrelated within a given fuel.

To quantify the effects of fuel composition modifications, gasoline with reduced aromatic content was used to determine the possible benzene reduction, and modified diesel fuel was used to determine the possible particulate matter emission reduction potential on diesel passenger cars and heavy-duty vehicles in the Federal Republic of Germany.

TABLE 12. PRESENT AND FUTURE LIMIT VALUES AND 1995 AVERAGE MARKET VALUES FOR GASOLINE FUEL IN THE FEDERAL REPUBLIC OF GERMANY

	Present Standards DIN EN 228	Market Average Germany 1995			Guidelines		Demand of ACEA
		Super Plus	Super	Regular	From 2000 Onward	From 2005 Onward	
RVP [kPa]	<70	66	65	65	<60	-	55–70*
Aromatics [vol.%]	-	43	38	31	<42	<35	<35
Benzene [vol.%]	<5	2.0	2.2	1.7	<1	-	<1
Oxygen [wt.%]	<2.8	1.3	0.4	0.1	<2.7	-	<2.7
Olefins [vol.%]	-	4	10	18	<18	-	<10
E 100 [vol.%]	40–70	49	51	53	>46	-	50–60
E 150 [vol.%]	-	84.5	84	84	>75	-	>85 (E 180)
Sulfur [wt.%]	<0.05	0.0065	0.0135	0.0275	<0.015	<0.005	<0.003
Lead [g/l]	< 0.013	0.001	0.001	0.001	<0.005		<0.005

* Valid for summer fuel

TABLE 13. PRESENT AND FUTURE LIMIT VALUES AND 1995 AVERAGE MARKET VALUES FOR DIESEL FUEL IN THE FEDERAL REPUBLIC OF GERMANY

	Present Standards DIN EN 590	Market Average Germany 1995	Guidelines		Demand of ACEA
			From 2000 Onward	From 2005 Onward	
Cetane Number	>49	52	>51	-	>55
Density [kg/m^3]	820–860	841	<845	-	<840
PAH [wt.%]	-	5.6	<11	-	<1
Total Aromatics [wt.-%]	-	-	-	-	<10
T 95 [°C]	<370	367	<360	-	<340
Sulfur [%m/m]	<0.05	0.04	>0.035	<0.005	<0.003

TABLE 14. EFFECT OF GASOLINE FUEL SPECIFICATIONS ON THE EMISSIONS OF PASSENGER CARS WITH AND WITHOUT CATALYSTS (0...NO EFFECT; ±0...-2 UP TO +2%; ↓ OR ↑...2 UP TO 10%; ↓↓ OR ↑↑...10 UP TO 20%; ↓↓↓ OR ↑↑↑...>20% EFFECT) [111–114]

Specification	Modification	Vehicles	CO	$C_{emiss.}$	$HC_{evap.}$	NO_x	Benzene	Butadiene	Aldehydes
Addition of Oxygenates	0 → 2.7%	w/o cat with cat	↓↓↓ ↓↓	↓ ↓	+0 +0	±0 +0	0 -0	0 0	↑↑ ↑
Reduction of Aromatics	40 → 25 vol.%	w/o cat with cat	↓ ↓	↓ ↓	0 0	↓ ±0	↓↓ ↓↓↓	0 +0	↑ ↑
Reduction of Benzene	3 → 2 vol.%	w/o cat with cat	0 0	0 0	-0 -0	0 0	↓↓ ↓↓	0 0	0 0
Reduction of Olefins	10 → 5 vol.%	w/o cat with cat	±0 0	↑ +0	-0 -0	↓ -0	0 0	↓↓ ↓↓	0 0

TABLE 14. (cont.)

Specification	Modification	Vehicles	CO	$C_{emiss.}$	$HC_{evap.}$	NO_x	Benzene	Butadiene	Aldehydes
Reduction of Sulfur	300 → 100 ppm	w/o cat with cat	0 ↓	0 ↓	0 0	0 ↓	0 ↓	0 ↓	0 ↑
Reduction of RVP	70 → 60 kPa	w/o cat with cat	0 0	±0 -0	↓↓↓ ↓↓	0 0	0 0	0 0	0 0
Increase of E 100	50 → 60%	w/o cat with cat	+0 +0	↓ ↓	±0 0	0 0	0 0	0 0	0 0
Increase of E 150	85 → 90%	w/o cat with cat	0 0	↓↓ ↓↓	0 0	↑ ↑	0 0	↓ ↓	↓ ↓

TABLE 15. EFFECT OF DIESEL FUEL SPECIFICATIONS ON THE EMISSIONS OF VEHICLES (0...NO EFFECT; ±0...-2 UP TO +2%; ↓ OR ↑...2 UP TO 10%; ↓↓ OR ↑↑...10 UP TO 20%; ↓↓↓ OR ↑↑↑...>20% EFFECT) [109, 112–120, 135–137]

Specification	Modification	Vehicles	CO	$HC_{emiss.}$	NO_x	Particles
Reduction of Sulfur	0.05 → 0.02 wt.%	PC[1)], LCV[2)] HCV[3)]	**0** **0**	0 0	0 0	↓ ↓
Reduction of Density	855 → 828 kg/m^3	PC, LCV HCV	↓↓ ↑	↓↓ ↑	↓↓ ↓	↓↓↓ ↓
Reduction of Polyaromatics	8 → 1 wt.%	PC, LCV HCV	+0 0	+0 -0	↓ ↓	↓↓ ↓
Reduction of Total Aromatics	30 → 10 wt.%	PC, LCV HCV	0 -0	0 -0	↓ ↓	↓ ↓
Increase of Cetane Number	50→55	PC, LCV HCV	↓↓ ↓↓	↓↓ ↓	±0 -0	±0 0
Reduction of T95	370→325°C	PC, LCV HCV	-0 ↑	+0 ↑	+0 -0	↓ -0

1. PC: Passenger Cars
2. LCV: Light Commercial Vehicles
3. HCV: Heavy Commercial Vehicles

Benzene Emissions

The influence of aromatic and benzene content of gasoline on passenger car emissions has been calculated for three scenarios. The calculation started from a base scenario that reflects fuel-sales-weighted actual fuel composition of approximately 40 vol.% aromatics and 2 vol.% benzene in Germany up to the year 1996, and was assumed to remain constant thereafter until the year 2020. The second scenario starts from an aromatic content of 35 vol.% and a benzene content of 1 vol.% as of the year 2000. Scenario 3 describes a fuel with an extremely low aromatic content, having a benzene content of 0.5 vol.% and remaining aromatics of 20 vol.%. The results of these calculations can be seen in Figure 68.

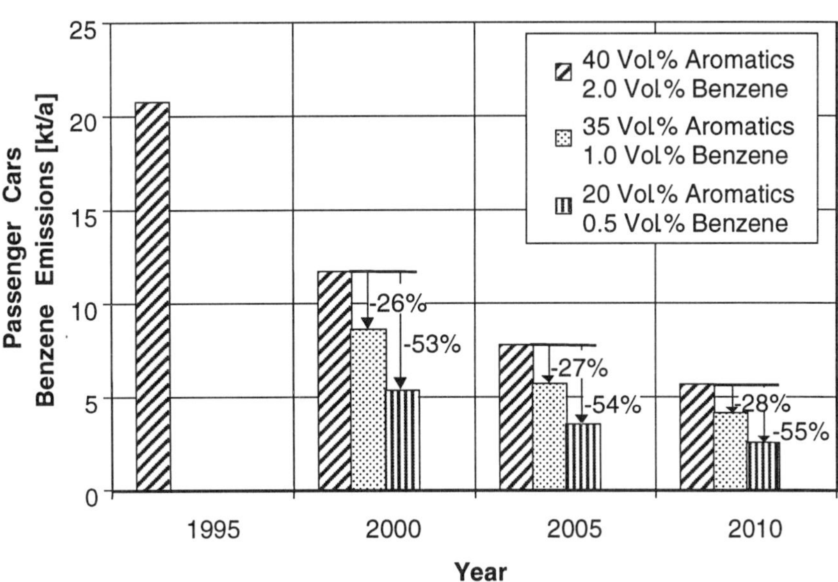

Figure 68. Effect of aromatic and benzene content variation in gasoline fuel on passenger car benzene emissions in the Federal Republic of Germany.

Concerning discussions about improved fuel qualities, it must be emphasized that their calculated emission reduction potential would lead to an immediate local and regional benefit as soon as such improved fuels were made available areawide by the oil industry.

Particulate Matter Emissions

The following scenario for particulate emission assumes fuel modifications as of the year 2000, as follows:

- Density reduced from 841 to 828 kg/m^3
- Polyaromatic content reduced from 8 to 1 wt.%
- Sulfur content reduced from 400 to 50 ppm
- Temperature T_{95} reduced from 367 to 325°C (693 to 617°F)
- Cetane number increased from 52 to 58

As shown in Figures 69 and 70, this fuel achieved a PM reduction of maximum 20% for passenger cars and maximum 10% for heavy-duty vehicles.

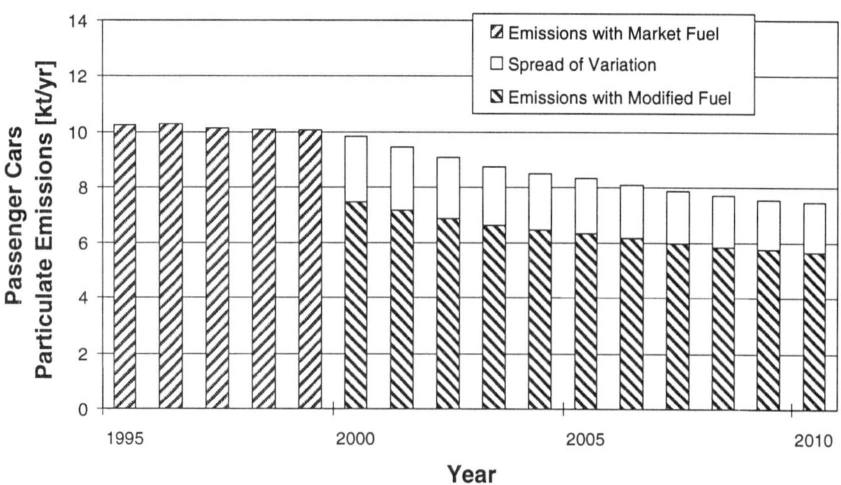

Figure 69. Effect of diesel fuel specification variation on passenger car particulate matter emissions in the Federal Republic of Germany.

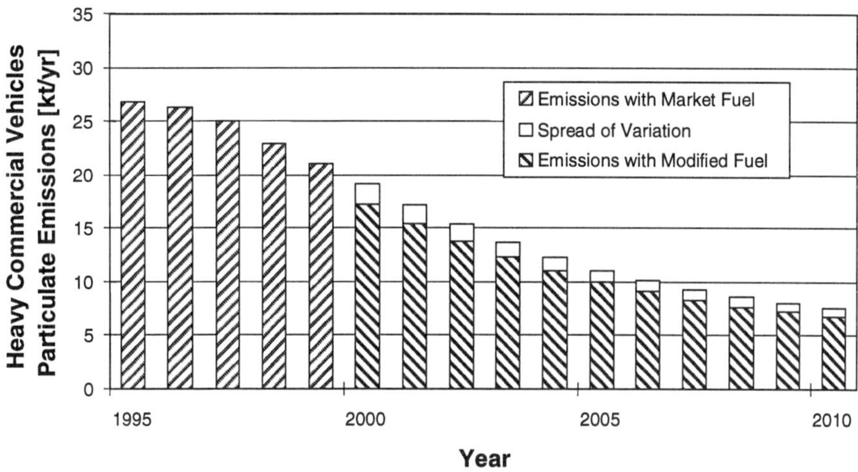

Figure 70. Effect of diesel fuel specification variation on heavy-duty vehicle particulate matter emissions in the Federal Republic of Germany.

5.2.4 Testing of In-Use Vehicles

The full benefit of the emission reduction potential of modern emission control technologies can materialize only if systems maintain their designed capabilities over an extended lifetime under real-world operating conditions. Several studies [121–124] have shown that approximately 1 to 2% of the vehicles in the field which are equipped with three-way catalysts had defects on parts of their emission control system, causing increased emissions.

As explained in Chapter 5, Section 5.2.2, continuously improved emission control technologies have been applied, in parallel to the introduction of more stringent legal requirements. These have not only reduced drastically the emission level but also have simultaneously improved the long-term emission characteristics of the vehicle in practical use.

To check and ensure this emission durability, legislation has established provisions for the periodic inspection of all emission-relevant components and systems. In the course of technical progress, it is expected that the effort necessary for such test methods can be reduced further through a transition toward self-diagnostic systems in the individual vehicle (onboard diagnosis

[OBD]), which would simultaneously achieve a substantial increase in efficiency of the program through the continuous surveillance. Such a system would automatically ensure the efficiency of all vehicles in the field at any time and under all operating conditions.

5.3 Vehicle User

The vehicle user can contribute substantially to reducing emissions and fuel consumption of his or her car by optimizing driving behavior. The following study [125] shows the effect of "adaptive" driving—a smooth driving style that avoids sharp accelerations and decelerations—on fuel consumption. Six vehicles with three-way catalyst technology were tested under urban and extra-urban (rural) driving conditions. The three types of driving behavior—namely, aggressive, normal, and smooth—were defined via average vehicle acceleration with 0.85–1.10 m/s^2, 0.65–0.80 m/s^2, and 0.46–0.65 m/s^2 and maximum vehicle acceleration with 3.0, 2.6, and 2.2 m/s^2, respectively.

Figures 71 through 74 show the results of the study. Urban driving was done either with cold start or warm start; extra-urban driving was always done with warm start. Emissions and fuel consumption were measured on all six vehicles for normal and aggressive driving; emissions under smooth driving were measured on only one vehicle.

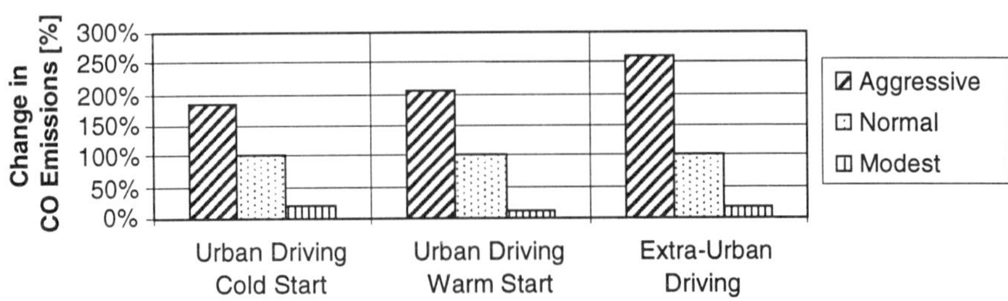

Figure 71. Effect of driving habits on carbon monoxide (CO) emissions.

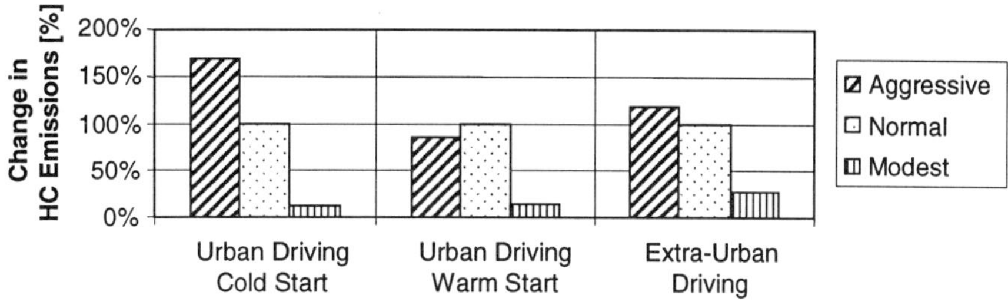

Figure 72. Effect of driving habits on hydrocarbon (HC) emissions.

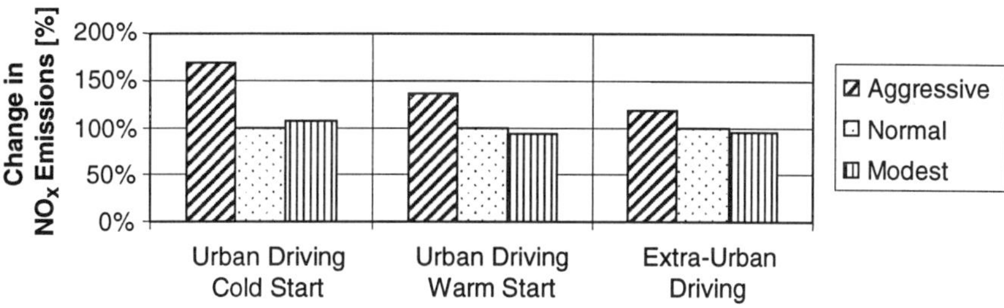

Figure 73. Effect of driving habits on nitrogen oxide (NO_x) emissions.

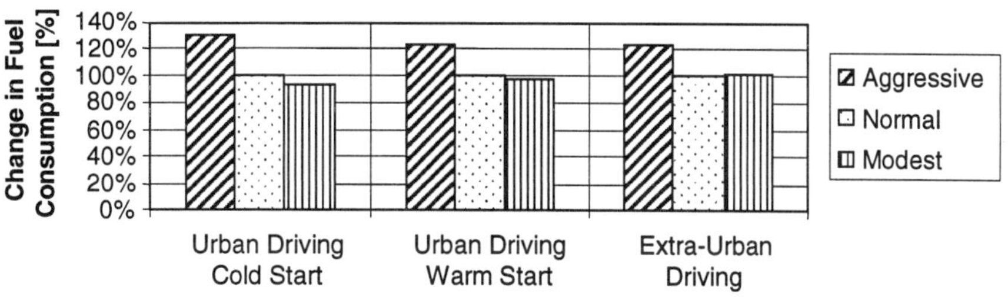

Figure 74. Effect of driving habits on fuel consumption.

5.4 Traffic Management

Among the so-called "non-technical measures," traffic management is gaining increased importance as an effective and efficient means to reduce fuel consumption and exhaust emissions. This is particularly true in cases where technical measures (i.e., emission control technologies in vehicles or fuel improvements in the market) may be insufficient to obtain necessary air quality targets.

One such cost-efficient traffic management measure is known as the "intelligent road." This means the establishment of a flexible speed limit that automatically adapts to existing traffic conditions. Such a system has been installed and tested on the Federal Highway (Bundesautobahn) A9 north of Munich [126]. The main reason for establishing this system was to reduce the number and severity of traffic accidents; however, it also showed the positive effect of reducing emissions and fuel consumption.

The reduction in emissions resulted primarily from a 90% elimination of stop-and-go events and from the elimination of the inefficient low speed ranges of 0 to 30 km/h [0 to 19 mph] in favor of higher speeds. Figures 75 and 76 show the fuel savings realized and the reduction in CO_2, HC, CO, NO_x, and PM emissions. Emission values are related to a time window with high traffic volume during morning hours from 6:00 to 10:00 A.M.

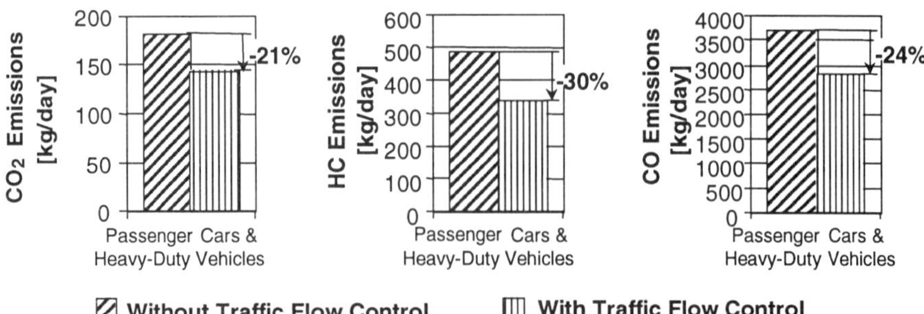

Figure 75. Carbon dioxide (CO_2), hydrocarbon (HC), and carbon monoxide (CO) emissions with and without traffic flow control on Federal Highway A9 from Highway Cross "Neufahrn" to Highway Connection "München Frankfurter Ring" from 6:00 to 10:00 A.M.

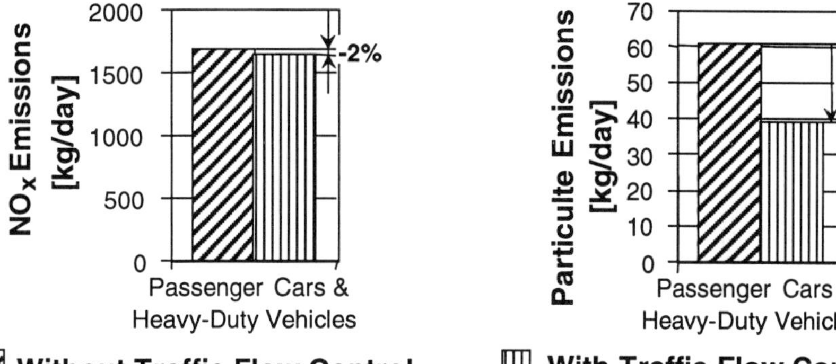

Figure 76. Nitrogen oxide (NO_x) and particulate matter (PM) emissions with and without traffic flow control on Federal Highway A9 from Highway Cross "Neufahrn" to Highway Connection "München Frankfurter R+ing" from 6:00 to 10:00 A.M.

The diagrams show a general reduction in fuel consumption (CO_2 emissions) and all exhaust gas constituents.

CHAPTER 6

Literature

[1] Brauer, H.: Handbuch des Umweltschutzes und der Umweltschutztechnik, Band 1: Emissionen und ihre Wirkungen, Springer-Verlag, Berlin, Heidelberg, New York, 1996.

[2] Metz, N., Grabbe, G.C.: Ozon—Ein vielschichtiges Problem, Studie der BMW AG, Abteilung Energie und Umwelt, München, Juni 1996.

[3] Wellburn, A.: Luftverschmutzung und Klimaänderung—Auswirkung auf Flora, Fauna und Mensch, Springer-Verlag, Berlin, Heidelberg, New York, ISBN 3-540-61831-7, Deutschland, 1997.

[4] Kuhn, M.: Klimaänderungen: Treibhauseffekt und Ozon. Hochschulschriftenreihen Forschung, Band 1, Kulturverlag Thaur/Tirol, 157 pp., 1990.

[5] Climate Change 1995, The Science of Climate Change, IPCC, Cambridge University Press, 1996.

[6] Umweltpolitik, Klimaschutz in Deutschland, Erster Bericht der Regierung der Bundesrepublik Deutschland nach dem Rahmenübereinkommen der Vereinten Nationen über Klimaänderungen, Bundesministerium für Umwelt, Naturschutz und Reaktorsicherheit, Bonn, September 1994.

[7] Lenz, H.P., Kohoutek, P., Pischinger, R., Hausberger, S.: Beeinflußungsmöglichkeiten des motorisierten Straßenverkehrs auf die CO_2-Emissionen, Int. Kongreß der AVL List GmbH, Graz am 24/25, August 1995.

[8] Heintz, A., Reinhardt, G.: Chemie und Umwelt, 2. Auflage, Vieweg & Sohn Verlagsgesellschaft, Braunschweig, Wiesbaden, ISBN 3-528-16349-6, Heidelberg, Deutschland, 1991.

[9] Dt. Bundestag—12. Wahlperiode: Erster Bericht der Enquete-Kommission “Schutz der Erdatmosphäre” zum Thema Klimaänderung gefährdet globale Entwicklung, Zukunft sichern—Jetzt handeln, Drucksache 12/2400, 31.3.1992.

[10] Hansen, J.E., Lebedeff, S.: Global Surface Air Temperatures: Update through 1987, Geophys. Res. Lett., 15, pp. 323–326, 1988.

[11] Ramanathan, V., Cicerone, R.J., Singh, B., Kiehl, J.T.: Trace Gas Trends and Their Potential Role in Climate Change, Geophys. Res. Lett., 90, pp. 5547–5566, 1985.

[12] Supplementary Diagrams from the Policymakers Summary of IPCC Working Group 1. Intergovernmental Panel on Climate, 1990.

[13] Lesch, K.-H., Cerveny, M., Leitner, A., Berger, B.: Treibhauseffekt Ursachen, Konsequenzen, Strategien, Umweltbundesamt Wien, Monographien Bd. 23, Juni 1990.

[14] Stahl, W., Berner, U.: Die Rolle der BRG in der Klimadiskussion, gehalten am 20 08.1997 in der Bundesanstalt für Geowissenschaften und Rohstoffe BGR, Hannover,1997.

[15] Rhode, H.: A Comparison of the Contributions of Various Gases to the Greenhouse Effect, Science, Vol. 248, Reports, pp. 1,217–1,219, June 1990.

[16] MacKenzie, J.J., Walsh, M.P.: Driving Forces: Motor Vehicle Trends and Their Implications for Global Warming, Energy Strategies, and Transportation Planning, WRI World Resources Institute, Washington, USA, December 1990.

[17] Hantel, M., Haslinger, A.: Klima—in Umweltwissenschaftliche Grundlagen und Zielsetzungen im Rahmen des Nationalen Umweltplans für die Bereiche Klima, Luft, Lärm und Geruch, Bundesministerium für Jugend, Umwelt und Familie, Schriftenreihe der Sektion I, Band 17/1994, erstellt durch die Österreichische Akademie der Wissenschaften, Wien, November 1994.

[18] Trenberth, K.E., Hurrell, J.W.: How Accurate Are Satellite "Thermometers"? Nature, Vol. 389, September 25, 1997.

[19] Singer, F.: The Scientific Case Against the Global Climate Treaty, Vortrag und Podiumsdiskussion im österreichischen Parlament am 3 November 1997.

[20] Treibhauseffekt, Ursachen, Konsequenzen, Strategien. Umweltbundesamt Wien, Monographien Bd. 23, Wien, Juni 1990.

[21] CDIAC: Carbon Dioxide Information Analysis Center, Meßwerte auf Datenträger, Oak Ridge, TN, 1997.

[22] Fossile Energieträger. Energie und Klima, Herausgegeben von der Enquete-Kommission "Vorsorge zum Schutz der Erdatmosphäre" des Deutschen Bundestages. Economica Verlag, Verlag C.F. Müller, 1990.

[23] Schönwiese, C.-D., Bissolli, P., Birrong, W., Ullrich, R.: Anthropogene, klimawirksame Spurengase, Mengen, Wirkung, Folgen, Gegenmaßnahmen, Klimatologische Aspekte, Berichte des Instituts für Meteorologie und Geophysik der Universität Frankfurt/Main, Nr. 87, 1990.

[24] Organisation for Economic Co-Operation and Development: OECD Environmental Data Compendium, 1993.

[25] Krapfenbauer, A., Wriessnig, K.: Anthropogene Umweltbelastung—Die Rolle der Landbewirtschaftung, Sonderdruck—Die Bodenkultur Journal für landwirtschaftliche Forschung, Band 46, Heft 3, Seite 269 ff., August 1995.

[26] Bolle, H.J.: Treibhauseffekt. Vortrag bei der Österreichischen Gesellschaft für Erdölwissenschaft, Schwechat, 18 April 1991.

[27] Heimann, M., Maier-Reimer, E.: On the Relations Between the Oceanic Uptake of CO_2 and Its Carbon Isotopes, Global Biogeochemical Cycles, Vol. 10, No. 1, pp. 89–110, March 1996.

[28] Walsh, M.P.: Motor Vehicle Pollution Control: A Global Overview, XXV FISITA Congress, 17–21 October, 1994, Beijing, SAE Paper 945112, Society of Automotive Engineers, Warrendale, PA, 1994.

[29] Woodwell, G.M.: Das Kohlendioxid-Problem, in Atmosphäre, Klima, Umwelt, 2. Auflage, Spektrum Akademischer Verlag, 1996.

[30] Faber, M., Jöst, F., Proops, J., Wagenals, G.: Wirtschaftliche Aspekte des Kohlendioxid-Problems, in Atmosphäre, Klima, Umwelt, 2. Auflage, Spektrum Akademischer Verlag, 1996.

[31] World Resources 1994/1995—A Guide to the Global Environment, United Nations Environment Programme, World Resources Institute, 4/1995.

[32] World Resources 1996/1997—A Guide to the Global Environment, United Nations Environment Programme, World Resources Institute, 12/1997.

[33] World Energy Outlook, Head Economic Analysis Division, OECD/IEA, Paris, 1994.

[34] International Energy Outlook, Energy Information Administration (EIA), Washington, USA, 1997.

[35] Environmental Data, Compendium 1995, OECD Organisation for Economic Cooperation and Development, Paris, 1995.

[36] Environmental Data, Compendium 1997, OECD Organisation for Economic Cooperation and Development, Paris, 1997.

[37] Teufel, D., Bauer, P., Braunfeld, S., Kilian, G., Wagner, Th.: Folgen einer globalen Motorisierung, Umwelt- und Prognose-Institut Heidelberg e.V., UPI-Bericht Nr. 35, März 1995.

[38] Orthofer, R.: Flüchtige halogenierte Kohlenwasserstoffe: Eine Abschätzung der Emissionssituation in Österreich, Österreichisches Forschungszentrum Seibersdorf, OEFZS-4526, NU-123/90, Jänner 1990.

[39] Shine, K., Derwent, R.G., Wuebbles, D.J., Morcrette, J.J.: Radiative Forcing of Climate, in "Bestandsaufnahme anthropogene Klimaänderungen: Mögliche Auswirkungen auf Österreich." Österreichische Akademie der Wissenschaften—Kommission für Reinhaltung der Luft, Verlag der ÖAW, 1990.

[40] Lenz, H.P., Kohoutek, P.: Weiterführende Literaturpflege des durchgeführten NO_x-Projektes, B2066, unveröffentlicht, Dez. 1993

[41] Fisher, D.A., Midgley, P.M.: Uncertainties in the Calculation of Atmospheric Releases of Chlorofluorocarbons, Journal of Geophysical Research, Vol. 99, No. D8, pp. 16,643–16,650, August 1994.

[42] Production, Sales, and Atmospheric Release of Fluorocarbons Through 1995, AFEAS, USA, Washington, DC, 1997.

[43] Midgley, P.M., McCulloch, A.: Estimated National Release to the Atmosphere of Chlorodifluoromethane (HCFC 22) During 1990, Pergamon, Atmospheric Environment, Vol. 31, No.6, pp. 809–811, Great Britain, 1997.

[44] Fraser, P., Cunnold, D., Alyea, F., Weiss, R., Prinn, R., Simmonds, P., Miller, B., Langenfelds, R.: Lifetime and Emission Estimates of 1,1,2-Trichlorotrifluorethane (CFC 113) from Daily Global Background Observations June 1982–June 1994, Journal of Geophysical Research, Vol. 101, No. C7, pp. 12,585–12,599, May 1996.

[45] Montrealer Protokoll—über Stoffe, die zu einem Abbau der Ozonschicht führen, Montreal, 16 September 1987.

[46] AFEAS, Production, Sales, and Athmospheric Release of Fluorocarbons Through 1995, Update 1997, Washington, DC, 1997.

[47] Lenz, H.P., Kohoutek, P.: Emissionen und Immissionen von nicht limitierten Abgaskomponenten, B2065, unveröffentlicht, Dez. 1993.

[48] United States Environmental Protection Agency, Presentation: Climate Change and Impacts, 1997.

[49] Streit, B.: 1994: Lexikon Ökotoxikologie, Zweite, aktualisierte und erweiterte Auflage, ISBN 3-527-30053-8, VCH Verlagsgesellschaft, Weinheim, Deutschland, 1994.

[50] Wei Min Hao, Ward, D.E.: Methane Production from Global Biomass Burning, Journal of Geophysical Research, Vol. 98, No. D11, pp. 20,657–20,661, November 1993.

[51] Crutzen, P.J.: Methane's Sinks and Sources, Nature, Band 350, S. 380–381, 1991.

[52] Hein, R., Crutzen, P.J., Heimann, M.: An Inverse Modeling Approach to Investigate the Global Atmospheric Methane Cycle, Global Biogeochemical Cycles, Vol. 11, No. 1, pp. 43–76, March 1997.

[53] Müller, J.-F., Brasseur, G.: Images: A Three-Dimensional Chemical Transport Model of the Global Troposphere, Journal of Geophysical Research, Vol. 100, No. D8, pp. 16,445–16,490, August 1995.

[54] Anastasi, C., Simpson, V.J.: Future Methane Emissions from Animals, Journal of Geophysical Research, Vol. 98, D4, pp. 7,181–7,186, April 1993.

[55] Mosier, A.R., Schimel, D.S.: Influence of Agricultural Nitrogen on Atmospheric Methane and Nitrous Oxide, Chemistry and Industry, 2 December 1991.

[56] Batjes, N.H., Bridges, E.M.: Potential Emissions of Radiatively Active Gases from Soil to Atmosphere with Special Reference to Methane: Development of a Global Database, Journal of Geophysical Research, Vol. 99, No. D8, pp. 16,479–16,489, August 1994.

[57] Anastasi, C., Dowding, M., Simpson, V.J.: Future CH_4 Emissions from Rice Production, Journal of Geophysical Research, Vol. 10, No. 1, pp. 89–110, March 1996.

[58] Law, K.S., Nisbet, E.G.: Sensitivity of the CH_4 Growth Rate to Changes in CH_4 Emissions from National Gas and Coal, Journal of Geophysical Research, Vol. 101, No. D9, pp. 14,387–14,397, June 1996.

[59] Lassey, K.R., Lowe, D.C.: A Source Inventory of Atmospheric Methane in New Zealand and Its Global Perspective, Journal of Geophysical Research, Vol. 97, No. D4, pp. 3,751–3765, March 1992.

[60] Lenz, H.P., Kohoutek, P.: Bedeutung der Stickstoffoxide und Emissionen und Immissionen von nicht limitierten Abgaskomponenten, B2134, unveröffentlicht, Dez. 1994.

[61] Elkins, J.W.: State of the Research for Atmospheric Nitrous Oxide (N_2O) in 1989. Contribution for the Intergovernmental Panel on Climate Change (IPCC), 1989.

[62] Bouwmann *et al.*: Uncertainties in the Global Source Distribution of Nitrous Oxide, Journal of Geophysical Research, Vol. 100, No. D2, pp. 2,785–2,800, February 1995.

[63] Potter, Ch.S. *et al.*: Process Modeling of Controls on Nitrogen Trace Gas Emissions from Soils Worldwide, Journal of Geophysical Research, Vol. 101, No. D1, pp. 1,361–1,377, January 1996.

[64] Khalil, M.A.K., Rasmussen, R.A.: The Global Sources of Nitrous Oxide, Journal of Geophysical Research, Vol. 97, No. D13, pp. 14,651–14,660, September 1992.

[65] Bange, H.W., Rapsomanikis, S, Andreae, M.O.: Nitrous Oxide in Coastal Waters, Global Biogeochemical Cycles, Vol. 10, No. 1, pp. 167–207, March 1996.

[66] Greenhouse Gas Emissions, The Energy Dimension. Heat of Publication Service, OECD, Paris, 1991.

[67] Fabian, P.: Atmosphäre und Umwelt, Berlin, Heidelberg, New York, Tokyo, Springer-Verlag, ISBN 3-540-55773-3, Deutschland, 1992.

[68] Köth-Jahr, I., Köllner, B.: Eine Bilanz neunjähriger Waldschadensforschung im Land Nordrhein-Westfalen - Ziele Ergebnisse Schlußfolgerungen, Forschungsschwerpunkt Luftverunreinigungen und Waldschäden, Ministerium f. Umwelt, Raumordnung und Landwirtschaft Nordrhein-Westfalen, ISSN 0934-5124, Düsseldorf, Juni 1993.

[69] Daten zur Umwelt—Der Zustand der Umwelt in Deutschland Ausgabe 1997, Umweltbundesamt Berlin, Erich Schmidt Verlag, ISBN 3-503-04310-1, Deutschland, 1997.

[70] Berge, E., Beck, J., Larssen, S., Moussiopoulos, N., Pulles, T.: Air Pollution in Europe 1997, EEA-European Environment Agency, ISBN 92-9167-059-6, Copenhagen, 1997.

[71] CORINAIR 1996.

[72] Stohl, A., Williams, E., Wotawa, G., Kromp-Kolb, H.: A European Inventory of Soil Nitric Oxide Emissions and the Effect of These Emissions on the Photochemical Formation of Ozone, Pergamon, Atmospheric Environment, Vol. 30, No. 22, pp. 3,741–3,755, Great Britain, 1996.

[73] Simpson, D., Guenther, A., Hewitt, C.N., Steinbrecher, R.: Biogenic Emissions in Europe, 1. Estimates and Uncertainties, Journal of Geophysical Research, Vol. 100, No. D11, Seiten 22,875–22,890, November 20, 1995.

[74] Chameides, W.L., Kasibhatala, R.S.,Yienger, J., Levy, H.: Growth of Continental-Scale Metro-Argo-Plexes, Regional Ozone Pollution and World Food Production Science 264, Seiten 74–76, 1994.

[75] Logan, J.A.: Nitrogen Oxides in the Troposphere: Global and Regional Budgets, Journal of Geophysical Research, Vol. 88, No. C15, pp. 10,785–10.807, December 20, 1983.

[76] Lee, D.S, *et. al.*: Estimates of Global NO_x Emissions and Their Uncertainties, Atmospheric Environment, Vol. 31, No. 12, pp. 1,735–1,749, 1997.

[77] Luftbelastung 1996, Meßresultate des Nationalen Beobachtungsnetzes für Luftfremdstoffe (NABEL), Schriftenreihe Umwelt Nr. 286, Bundesamt für Umwelt, Wald und Landschaft, Bern, 1997.

[78] Schermann, G.: Monatsmittelwerte 1996, Luftmeßnetz der MA 22, Persönliche Mitteilung, 1998.

[79] Redl, M., Lorenz, D.: Luftmeßnetz Jahresbericht 1996: Amt der OÖ Landesregierung Meßberichte, 1997.

[80] Luftqualität in Nordrhein-Westfalen, TEMES- und LIMES-Jahresberichte, Landesanstalt für Immissionsschutz Nordrhein-Westfalen, Essen, 1993, 1994, 1995, 1996, 1997.

[81] Kohoutek, P.: Entwicklung einer Datenbank zur Berechnung von Abgasemissionen benzinbetriebener PKW unter besonderer Berücksichtigung der Kraftstoffzusammensetzung, VDI-Fortschritt-Berichte, Reihe 12 Nr. 275, Düsseldorf, 1996.

[82] Balrutsch, M., Hanewald, K., Siegmund, A., Stec-Lazaj, W., Wunderlich, W.: Lufthygienischer Jahresbericht 1994, Hessische Landesanstalt für Umwelt, 1994 und persönliche Mitteilung.

[83] Immissionsmeßwerte Landesumweltamt Bayern, persönliche Mitteilung.

[84] Baum, F.: Luftreinhaltung in der Praxis, Oldenburg-Verlag, Wien-München, 1988.

[85] Lipkea, W.H., Johnson, J.H., Vuk, C.T.: The Physical and Chemical Character of Diesel Particulate Emissions—Measurement Techniques and Fundamental Considerations, Society of Automotive Engineers, SAE Paper 780108, Society of Automotive Engineers, Warrendale, PA, 1978.

[86] Eisert, M.: Partikel-Emissionen und Immissionen, Diplomarbeit TU Wien, Juli 1998.

[87] Mandl, M.: Messung von Benzol, PAH, Dieselruß, an einer verkehrsrelevanten Meßstelle in der Stadt Salzburg, Schlußfolgerungen für das Salzburger Meßnetz im Hinblick auf Parameter und Meßmethoden, Diplomarbeit TU Graz, December 1995.

[88] Benzol, BUA-Stoffbericht 24. Beratergremium für umweltrelevante Altstoffe (BUA) der Gesellschaft deutscher Chemiker. Weinheim, Basel (Schweiz), Cambridge, New York, 1988.

[89] Diverse Monats und Jahresberichte der Landesanstalt für Umweltschutz Baden-Württemberg erarbeitet durch die Gesellschaft für Umweltmessungen und Umwelterhebungen (UMEG).

[90] Directive on Fuel Quality, European Automobile Manufacturers Association (ACEA), Emission 2000, No. 165, Brussels, 9 July 1998.

[91] Jahresbericht 1996 des Umweltbundesamtes Berlin.

[92] Spiecker, H., Köhl, M., Mielikäinen, K., Skovsgaard, J.P.: Growth Trends in European Forests, Springer-Verlag, Heidelberg, 1996.

[93] Landgrebe, J.: Statement des Umweltbundesamtes Berlin, Gasrückführung beim Betanken von Fahrzeugen, Tagung am 3. November 1993, Seite 93–98, Umweltbundesamt Wien, 1994.

[94] LAI—Länderausschuß für Immissionsschutz: Krebsrisiko durch Luftverunreinigungen, Entwicklung von "Beurteilungsmaßstäben für kanzerogene Luftverunreinigungen" im Auftrag der Umweltministerkonferenz. Herausgegeben vom Ministerium für Umwelt, Raumordnung und Landwirtschaft des Landes Nordrhein-Westfalen, Düsseldorf, 1992.

[95] Umweltbundesamt Berlin, Immissionsmeßwerte auf Datenträger.

[96] Orthofer, R., Vesely, A.: Abschätzung von toxischen Emissionen (PCDD, PCDF, PAH, BaP) aus Verbrennungsprozessen in Österreich, Österreichisches Forschungszentrum Seibersdorf, Nachdruck zu OEFZS-A-1779/August 1990, October 1990.

[97] Funcke, W., König, J., Balfanz, E., Romanowski, Th., Großman, I. 1982: Analyse von polyzyklischen aromatischen Kohlenwasserstoffen auf Immissionsstäuben aus dem Ruhrgebiet und einer ländlichen Kleinstadt, Staub—Reinhalt. Luft 42, Nr. 5, S. 192/97, 1982.

[98] Plaßmann, E., Waldeyer, H., Brosthaus, J.: Minderung der Luftbelastung und des Energiebedarfs durch die Verkehrsmittel im Vergleich, Verkehr in der Mitte Europas—Deutscher Ingenieurtag 1993, VDI-Berichte 1041, VDI-Verlag, Düsseldorf 1993.

[99] Motorisierung—Frauen geben Gas, Neue Techniken senken Verbrauch und Emissionen, Deutsche Shell AG, Shell PKW-Szenarien, Hamburg, 1997.

[100] Mobil bleiben Umwelt schonen, Energieprognose '94, Esso AG, Hamburg, 1995.

[101] Ratzenberger, R., Hild, R., Langmantel, E.: Vorausschätzung der Verkehrsentwicklung in Deutschland bis zum Jahr 2010, ifo Institut für Wirtschaftsforschung Abt. Verkehr, München, September 1995.

[102] Hassel, D., Jost, P., Weber, F.-J., Dursbeck, F., Sonnborn, K.-S., Plettau, D.: Abgas-Emissionsfaktoren von PKW in der Bundesrepublik Deutschland, Abgasemissionen von Fahrzeugen der Baujahre 1986 bis 1990. UFOPLAN-Nr. 104 05 152 und 104 05 509 UBA-FB 91-042, Abschlußbericht. TÜV-Rheinland, December 1993.

[103] Palm, I., Regniet, G., Schmidt, G.: Ermittlung der PKW- und NFZ-Jahresfahrleistung 1993 auf allen Straßen in der Bundesrepublik Deutschland. FE-Nr.: 90434/94, Dr.-Ing. H. Heusch—Dipl.—Ing. J. Boesefeldt, Aachen, 1996.

[104] Hausberger, S., Sams, T., Sturm, P.: KEMIS—Programm zur Simulation von streckenbezogenen Abgasemissionen des Straßenverkehrs, Programmbeschreibung. Institut für Verbrennungskraftmaschinen und Thermodynamik der TU Graz, 1994.

[105] Straßenverkehr in Baden-Württemberg—Jahresvergleich 1991/1990. Hrsg. Verkehrsministerium in Baden-Württemberg, Abt. Straßenbau, Stuttgart, November 1992.

[106] Directive on Emissions from Passenger Cars and Light-Duty Commercial Vehicles, European Automobile Manufacturers Association (ACEA), Emission 2000, No. 164, Brussels, 7 July 1998.

[107] World-Wide Fuel Charter, American Automobile Manufacturers Association (AAMA), European Automobile Manufacturers Association (ACEA), Japan Automobile Manufactures Association (JAMA), June 1998.

[108] Prüller, S.: Ermittlung von Partikel-Emissionen, Diplomarbeit TU Wien, unveröffentlicht, December 1997.

[109] Krumm, H., Lange, W.W., Holthusen, E.G.: Geringere Umweltbelastung durch verbesserte Fahrzeugbetriebsstoffe, Vortrag: Technische Akademie Esslingen 26/27 September 1996, Shell Technischer Dienst, Deutsche Shell AG, ISSN 0934-3601, Hamburg, 1996.

[110] Strand, A., Hov, O.: The Impact of Man-Made and Natural NO_x Emissions on Upper Tropospheric Ozone: A Two-Dimensional Model Study, Pergamon, Atmospheric Environment, Vol. 30, No. 8, pp. 1,291–1303, Great Britain, 1996.

[111] Barrett, E., Pollitt, T.: Auswirkung einer veränderten Benzinzusammensetzung in Europa, Bericht UBA-FB 94-028 Berlin, April 1994.

[112] Kempe, J.: 1. Internationales Kraftstoff-Kolloquium, Bericht über die Veranstaltung in Bad Nauheim am 26/27 September 1994, Erdöl Erdgas Kohle, 111 Jahrgang, Heft 3, S. 137 ff., März 1995.

[113] ACEA, Europia (Hg.), European Programme on Emissions, Fuels, and Technologies (EPEFE), Brussels, 1996.

[114] Reglitzky, A.A., Lange, W.W., Schnieder, N., Krumm, H.: Saubere Luft und Klimaschutz ein Zielkonflikt? Shell Technischer Dienst, Deutsche Shell AG, ISSN 0934-3601, Hamburg, November 1994.

[115] Lange, W.W., Schäfer, A., Le'Jeune, A., Naber, D., Reglitzky, A.A., Gairing, M.: Der Einfluß von Kraftstoffeigenschaften auf die Abgasemissionen Moderner Dieselmotoren von Mercedes-Benz, MTZ Motortechnische Zeitschrift 55, 1994.

[116] Gairing, M., Schäfer, A., Naber, D., Lange, W.W., Graupner, O., Stradling, R.: Einfluß von Polyaromaten, Schwefelgehalt und Viskosität auf die Abgasemissionen Moderner Mercedes-Benz-Dieselmotoren, MTZ Motortechnische Zeitschrift 58, 1997.

[117] Lange, W.W.: The Effect of Fuel Properties on Particulate Emissions in Heavy-Duty Truck Engines Under Transient Operating Conditions, SAE Paper 912425, Society of Automotive Engineers, Warrendale, PA. 1991.

[118] Spreen, K., Ullmann, T., Mason, R.L.: Effects of Cetane Number, Aromatics, and Oxygenates on Emissions from a 1994 Heavy-Duty Diesel Engine with Exhaust Catalyst, SAE Paper 950250, Society of Automotive Engineers, Warrendale, PA, 1995.

[119] Tsurutani, K., Takei, Y., Fujimoto, Y., Matsudaira, J., Kumamoto, M.: The Effects of Fuel Properties and Oxygenates on Diesel Exhaust Emissions, SAE Paper 952349, Society of Automotive Engineers, Warrendale, PA, 1995.

[120] Ullmann, T., Spreen, K.B., Mason, R.L.: Effects of Cetane Number, Cetane Improver, Aromatics, and Oxygenates on 1994 Heavy-Duty Diesel Engine Emissions, SAE Paper 941020, Society of Automotive Engineers, Warrendale, PA, 1994.

[121] Erhebungen des Straßenverkehrsamtes Zürich. Autotechnik, Nr. 9, S. 31, 1991.

[122] Merbaul, H.: Auswertung der jährlichen Überprüfung von im Verkehr befindlichen Kraftfahrzeugen. Diplomarbeit am Institut für Verbrennungskraftmaschinen und Kraftfahrzeugbau der TU Wien, Berichtsnummer B 1785 unveröffentlicht, November 1990.

[123] Marek, T.: Möglichkeiten zur Detektierung von hochemittierenden Kraftfahrzeugen im Straßenverkehr. Diplomarbeit am Institut für Verbrennungskraftmaschinen und Kraftfahrzeugbau der TU Wien, Berichtsnummer B 2071 unveröffentlicht, Jänner 1994.

[124] Lenz, H.P., Kohoutek, P.: Schlußbericht über die "Überprüfung der derzeit im Verkehr befindlichen PKW mit Ottomotor und geregeltem Dreiwege-Katalysator," Institut für Verbrennungskraftmaschinen und Kraftfahrzeugbau der TU Wien, Berichtsnummer B 2030 unveröffentlicht, April 1993.

[125] Vlieger De I.: On-Board Emission and Fuel Consumption Measurement Campaign on Petrol-Driven Passenger Cars, Pergamon, Atmospheric Environment, Vol. 31, No. 22, pp. 3,753–3,761, Great Britain, 1997.

[126] Metz, N., Schlichter H., Schellenberg H.: Reduzierung des Kraftstoffverbrauchs, der CO_2- und Abgas-Emissionen durch die Linienbeeinflussungsanlage auf der Bundesautobahn A9, Umweltkongreß 1996 der Stadt Mannheim, der Technischen Akademie und des VDI, Mannheim, Juli 1996.

[127] Lenz, H.P., Pucher, E., Kohoutek, P., Rennenberg, H., Hahn, J., Elstner, E.F., Hippeli, S.: Emissionen, Immissionen und Wirkung von Abgaskomponenten, Forschrittsberichte-VDI, Reihe 12, Verkehrstechnik, Fahrzeugtechnik, Nr. 183, Düsseldorf, 1993.

[128] Graedel, T.E., Crutzen, P.J.: Veränderungen in der Atmosphäre, Spektrum der Wissenschaft, November 1989.

[129] Kasting, J.F., Toon, O.B., Pollak, J.B.: Die Entwicklung des Klimas auf den erdähnlichen Planeten, Spektrum der Wissenschaft, April 1988.

[130] Broecker, W.S.: Plötzliche Klimawechsel, Spektrum der Wissenschaft, Jänner 1996.

[131] Luftgüte-Meßnetz BLUME, Senatsverwaltung für Stadtentwicklung, Umweltschutz und Technologie, diverse Jahresberichte Luftgütemeßdaten.

[132] Dobson, G.: Exploring the Atmosphere, Clarendon Press, Oxford, 1968.

[133] Marx, T.: Rückgewinnung gasförmiger Anästhetika—Xenon als Narkosegas, Forschungsgruppe Anästhesie und Umwelt/Xenon, 1998.

[134] Berdowski, J.J.M., Mulder, W., Veldt, C., Visschedijk, A.J.H., Zandveld, P.Y.J.: Particulate Matter Emissions (PM_{10}–$PM_{2.5}$—$PM_{0.1}$) in Europe in 1990 and 1993, TNO-Report R96/472, February 1997.

[135] Kleinschek, G., Richter, M, Röj, A., Singer, M., Stein, H.J.: Influence of Diesel Fuel Quality on Heavy-Duty Diesel Engine Emissions, Further Investigation on the EPEFE Results by the ACEA Heavy-Duty Diesel Truck Manufacturers, ACEA Report, March 1997.

[136] Lange, W.W., Reglitzky, A., Gadd, P., Schögl, H., Richter, K., Zürner, H.J.: Das Potential der Kraftstoffqualität auf Verbrennung und Abgas-emission von abgasoptimierten Dieselmotoren, 5. Aachener Kolloquium, 1995.

[137] Stradling, R., Gadd, P., Singer, M., Operti, C.: The Influence of Fuel Properties and Injection Timing on the Exhaust Emissions and Fuel Consumption on an Iveco Heavy-Duty Diesel Engine," SAE Paper 971636, Society of Automotive Engineers, Warrendale, PA, 1997.

[138] Kemper, G.: 23.Verordnung §40 BImSchG Absatz 2: Konzeption und Inhalte, 1. Düsseldorfer Umweltkonferenz, 20–21 Juni 1995, Sektion 2 Luftbelastung durch den Verkehr, Kommission Reinhaltung der Luft im VDI und DIN, Band 23, Düsseldorf, 1995.

[139] Kyoto Protocol to the United Nations Framework Convention on Climate Change, Kyoto, 1–10 December 1997.

[140] Kolar, J.: Stickoxide und Luftreinhaltung, Grundlagen, Emissionen, Transmissionen, Immissionen, Wirkungen, Springer-Verlag, Berlin, Heidelberg, New York, 1990.

[141] Schumann, U.: The Impact of Nitrogen Oxides Emissions from Aircraft upon the Atmosphere at Flight Altitudes—Results from the AERONOX Project, Pergamon, Atmospheric Environment, Vol. 31, No. 12, pp. 1,723–1,733, Great Britain, 1997.

[142] Köhler, I., Sausen, R., Reinberger, R.: Contributions of Aircraft Emissions to the Atmospheric NO_x Content, Pergamon, Atmospheric Environment, Vol. 31, No. 12, pp. 1,801–1,818, Great Britain, 1997.

[143] Lenz, H.P., Cozzarini, Ch.: Emissionen und Immissionen von Abgas-komponenten 1995, B2254, unveröffentlicht, Mai 1996.

[144] Lenz, H.P., Cozzarini, Ch.: Emissionen und Immissionen von Abgas-komponenten 1996, B2319, unveröffentlicht, Mai 1997.

[145] Lenz, H.P., Cozzarini, Ch.: Besprechungsunterlagen des AK-Emissionen Luftqualität des Jahres 1997 vom 23 Jänner 1997, 4 Juli 1997, 31 Oktober 1997, unveröffentlicht, 1997.

[146] Der Umweltbericht von Volkswagen, 1997.

[147] Aktuelles Lexikon, BMW AG, 1992.

[148] MAK- und BAT-Werte-Liste 1996. Deutsche Forschungsgemeinschaft, Senatskommission zur Prüfung gesundheitsschädlicher Arbeitsstoffe, Mitteilung 32. VCH Verlagsgesellschaft mbH, Weinheim, BRD, 1996.

Index

f = figure; t = table

Acid rain, forest damage and, 32
Air polluting substances
    atmospheric effects of, 3-6
        local and global, 5-6
    of long-life, 6
    path of, 3
    of short-life, 6
Air quality, 4
Airborne dust
    air quality and, 52-53, 55-58
    atmospheric concentrations of, 56-57*t*, 58*f*, 58-60, 59*f*
    size ranges of, 53*f*
    sources of, 52
Airborne dust emissions, 53-55, 55*f*
Anthropogenic greenhouse gases, 9*f*
Anthropogenic sources
    of airborne dust, 52
    of atmospheric trace substances, 4
Assembly line, exhaust emission standards for, 85*f*
Atmosphere, air polluting substances and, 3-6
    greenhouse effect, 7-12
    local and global, 5-6
    Montreal Protocol, 22-23
    stratospheric ozone depletion, 12-29
    water vapor, 13
Atmospheric concentrations
    of CFCs, 22*f*
    of methane, 23-25, 24*f*
    of nitric oxide, 39*t*
    of nitrogen dioxide, 38*t*
    of nitrogen oxides, 39*f*, 39-40, 40*f*
    of nitrous oxide, 25-29, 26*f*
    of non-methane hydrocarbons, 44*f*

Atmospheric concentrations *(continued)*
of sulfur dioxide, 49-50*t*, 50, 51*f*
of suspended particulate matter, 59*f*
Atmospheric ozone. *See* Stratospheric ozone
Atmospheric trace substances, sources of, 4

Benzene (C_6H_6), 68
air quality and, 69-70, 70*t*, 71*t*
atmospheric concentrations of, 70*t*, 71, 72*f*
Benzene emissions, 69, 69*f*
exhaust emission standards for, 93*f*, 93-94

Cancer-causing components, 71*t*
Carbon dioxide (CO_2)
atmospheric, 14-15
description of, 14
Carbon dioxide emissions
conflict between nitrogen oxide emissions and, 33*f*
development of, 15, 16*f*
prognosis for, 18, 18*f*
sources of, 15, 17*f*
traffic flow control and, 98*f*
trend of, 18, 19*f*
yearly global spreads of, 15, 17*f*
Carbon monoxide (CO)
air quality and, 63, 63*t*
atmospheric concentrations of, 63*t*, 63-64, 64*f*
description of, 60
Carbon monoxide emissions
driving habits and, 96*f*
in Europe, 60, 61*f*
exhaust emission standards for, 84*f*, 87*f*
in Federal Republic of Germany, 61, 62*f*
traffic flow control and, 98*f*
CFCs
emissions and concentrations in atmosphere for, 22*f*
ozone depletion potential and, 14*t*
see also Halogenated hydrocarbon emissions
CH_3Br. *See* Methylbromide

CH_3Cl. *See* Methylchloride
CO. *See* Carbon monoxide
CO_2. *See* Carbon dioxide
Commercial vehicles, exhaust emission standards for, 85*f*
Coronene, 73, 74*f*

Diesel fuel
particulate matter emissions and, 94*f*, 95*f*
specifications for, 91*t*, 92*t*
Driving habits, effect of
on carbon monoxide emissions, 96*f*
on fuel consumption, 97*f*
on hydrocarbon emissions, 97*f*
on nitrogen oxide emissions, 97*f*
Dust. *See* Airborne dust

Earth temperature, trend of, 9*f*
Emissions
of airborne dust, 53-55, 55*f*
of carbon dioxide, 14-19
development of, 16*f*
from fossil fuel burning, 18*f*
sources of, 17*f*
trend of, 19*f*
yearly, 17*f*
of CFC 11, 22*f*
evaluation of, 5
of greenhouse gases, 9*f*
comparison of, 11*t*
greenhouse effect, 7-12
potentials of, 12*f*
of halogenated hydrocarbons, 19-23, 21*f*
global warming potential and, 21*t*
of methane
development of, 24*f*
sources of, 25*f*
Montreal Protocol and, 22-23
of nitrous oxide
development of, 26*f*

Emissions *(continued)*
of nitrous oxide *(continued)*
sources of, 28*f*
spread of, 27*f*
from road traffic, 75, 76*f*
in Germany, 76-82
sources of, 4
stratospheric ozone depletion, 12-29
water vapor, 13
see also specific types of emissions
Euro 2 legislation, 84
Euro 3 legislation, 84, 87
Euro 4 legislation, 84, 85, 87
Europe
airborne dust emissions in, 53-54, 54*f*
carbon monoxide emissions in, 60, 61*f*
nitrogen oxide emissions in, 33, 35*f*, 35-36
non-methane hydrocarbon emissions in, 41, 41*f*, 42*f*
sulfur dioxide emissions in, 45*f*, 45-46, 46*f*
Exhaust emissions
components of
acid rain, 32
benzene, 68-72
carbon monoxide, 60-64
dust and particulate matter, 52-60
nitrogen oxides, 32-40
non-methane hydrocarbons, 40-45
ozone, 64-68
polycyclic aromatic hydrocarbons, 72-74
smog, 31
sulfur dioxide, 45-51
standards
for assembly line, 85*f*
for benzene, 93-94
for carbon monoxide, 87*f*
for commercial vehicles, 85*f*
effectiveness of, 85
fuel composition and, 89-95
for hydrocarbons, 86*f*
for in-use vehicles, 87-89, 95-96
level of, 83-87
for nitrogen oxide, 86*f*

Exhaust emissions *(continued)*
standards *(continued)*
for particulate matter, 94-95
for passenger cars, 84*f*

Federal Republic of Germany
airborne dust emissions in, 54-55, 55*f*, 56*f*
atmospheric concentrations of sulfur dioxide in, 49-50*t*, 51*f*
benzene emissions in, 69, 69*f*
carbon monoxide emissions in, 61, 62*f*
new passenger car registrations in, 77*f*
nitrogen oxide emissions in, 36, 36*f*, 37*f*
non-methane hydrocarbon emissions in, 42, 43*f*
passenger car life cycle probability in, 77*f*
road traffic emissions in, 76-82
sulfur dioxide emissions in, 46-47, 47*f*, 48*f*
suspended particulate matter concentrations in, 56-57*t*
Forest damage, acid rain and, 32
Fuel composition
exhaust emission standards for, 89-95
of gasoline fuel, 90*t*
Fuel consumption, driving habits and, 97*f*

Gasoline fuel, specifications for, 90*t*, 91-92*t*
Germany. *See* Federal Republic of Germany
Global effects, of long-life air polluting substances, 6
Global warming potential (GWP), halogenated hydrocarbon emisssions and, 21*t*
Greenhouse effect, 7-8, 10-11
earth temperature trend, 9*f*
greenhouse gases and, 9*f*
Kyoto Conference and, 8
Greenhouse gases
emissions, 11*t*
greenhouse effect and, 9*f*
potentials of, 12*f*
trace gases, lifetime, 11*t*
water vapor, 13
GWP. *See* Global warming potential

Halogenated hydrocarbon emissions, 20
  as carbon dioxide equivalents, 21*f*
  global warming potential and, 21*t*
  Montreal Protocol and, 22-23
  most probable values for, 21*t*
  see also CFCs
Halogenated hydrocarbons, 19-23
  characteristics of, 20
  definition of, 19
Halons, ozone depletion and, 14*t*
HCFCs, ozone depletion and, 14*t*
Heavy-duty vehicles, particulate matter emissions from, 95*f*
HFCs, ozone depletion and, 14*t*
Hydrocarbon emissions
  driving habits and, 97*f*
  effect of mileage and production year on, 89*f*
  from passenger cars, 86*f*, 88*f*
  traffic flow control and, 98*f*
Hydrocarbons
  exhaust emission standards for, 84*f*, 86*f*, 88*f*, 89*f*
  see also Halogenated hydrocarbons; Non-methane hydrocarbons; Polycyclic aromatic hydrocarbons

In-use vehicles
  emission standards for, 87-89
  testing of, 95-96

Kyoto Conference, greenhouse effect and, 8

Legislative measures
  for assembly line exhaust emissions, 85*f*
  for benzene emissions, 93-94
  for diesel fuel, 91*t*, 92*t*
  effectiveness of, 85
  for fuel composition, 89-95
  for gasoline fuel, 90*t*, 91-92*t*
  for heavy-duty vehicle emissions, 95*f*

Legislative measures *(continued)*
for in-use vehicles
emissions, 87-89
testing, 95-96
level of emission standards, 83-87
for particulate matter emissions, 94-95, 95*f*
for passenger car emissions, 84*f*
benzene, 93*f*
carbon monoxide, 87*f*
hydrocarbons, 86*f,* 88*f,* 89*f*
nitrogen oxide, 86*f*
particulate matter, 94*f*
Local effects, of short-life air polluting substances, 6
London smog, 31
Long-life components, with global effects, 6
Los Angeles smog, 31

Methane
atmospheric concentrations of, 24*f*
description of, 23
Methane emissions
global spreads for, 24*f*
sources of, 25*f*
Methylbromide (CH_3Br), 19
Methylchloride (CH_3Cl), 19
Montreal Protocol, halogenated hydrocarbon emisssions and, 22*f,* 22-23

N_2O. *See* Nitrous oxide
Natural sources
of airborne dust, 52
of atmospheric trace substances, 4
Nitric oxide, atmospheric concentrations of, 39*t*
Nitrogen dioxide, atmospheric concentrations of, 38*t,* 39*f*
Nitrogen oxide, 25
atmospheric concentrations of, 39-40, 40*f*
judgment criteria for, 37, 38*t,* 39*t*
Nitrogen oxide emissions
carbon dioxide emissions and, 33*f*
driving habits and, 97*f*

Nitrogen oxide emissions *(continued)*
in Europe, 33, 35*f,* 35-36
in Germany, 36, 36*f,* 37*f*
global yearly, 34*f*
standards for, 84*f,* 86*f*
traffic flow control and, 99*f*
Nitrous oxide (N_2O), 25-29
description of, 25-26
development of, 26, 26*f*
tropospheric, 26
Nitrous oxide emissions
passenger cars and, 28-29
sources of, 27-28, 28*f*
spreads of, 27, 27*f*
NMHC. *See* Non-methane hydrocarbons
Non-methane hydrocarbon emissions
in Europe, 41, 41*f,* 42*f*
in Germany, 42, 43*f*
Non-methane hydrocarbons (NMHC)
air quality and, 40-44
atmospheric concentrations of, 44*f*

ODP. *See* Ozone depletion potential
Ozone. *See* Stratospheric ozone
Ozone depletion potential (ODP), 14*t*

PAHs. *See* Polycyclic aromatic hydrocarbons
Particulate matter (PM)
size ranges of, 53*f*
see also Airborne dust
Particulate matter emissions
diesel fuel specifications and, 94*f,* 95*f*
fuel modifications for, 94
sources of, 52
standards for, 94-95
traffic flow control and, 99*f*
Passenger car emissions
benzene, 93*f,* 93-94
carbon monoxide, 87*f*

Passenger car emissions *(continued)*
gasoline fuel specifications and, 91-92*t*
hydrocarbons, 86*f*, 88*f*, 89*f*
nitrogen oxide, 86*f*
particulate matter, 94*f*
standards for, 83, 84*f*
PM. *See* Particulate matter
Polycyclic aromatic hydrocarbons (PAHs), 72-73
atmospheric concentrations of, 73, 74*f*

Road traffic emissions, 75, 76*f*
benzene, 93-94
calculation method for, 81*f*, 81-82
daily traffic volume and, 80*f*
diesel fuel specifications and, 91*t*, 92*t*
fuel composition and, 89-95
gasoline fuel specifications and, 90*t*, 91-92*t*
HC cold-start effect and, 80*f*
heavy-duty vehicles and, 79*f*, 95*f*
in-use vehicles and, 87-89, 95-96
legislative measures for
effectiveness of, 85
level of emission standards, 83-87
particulate matter, 94, 95*f*
passenger cars and, 84*f*
benzene emissions, 93*f*
carbon monoxide emissions, 87*f*
hydrocarbon emissions, 86*f*, 88*f*, 89*f*
life cycle probability, 77*f*
nitrogen oxide emissions, 86*f*
particulate matter emissions, 94*f*
population, 78*f*
new registrations, 77*f*
traffic flow control and, 98-99
vehicle user and, 96*f*, 97*f*

Short-life components, with local effects, 6
Smog, 31
Soot, atmospheric concentrations of, 59*f*, 59-60

SPM. *See* Suspended particulate matter
Stratospheric ozone, 64-65
 air quality and, 65, 66*t*
 atmospheric concentrations of, 66, 66*t*, 67*f*, 68
 depletion of
  carbon dioxide and, 14-19
  halogenated hydrocarbons and, 19-25
  nitrous oxide and, 25-29
  water vapor and, 13
Sulfur dioxide
 air quality and, 49-50*t*
 atmospheric concentrations of, 49-50*t*, 50, 51*f*
 impact of, 48
Sulfur dioxide emissions
 in Europe, 45*f*, 45-46, 46*f*
 in Federal Republic of Germany, 46-47, 47*f*, 48*f*
Summer smog, 83
Suspended particulate matter (SPM)
 atmospheric concentrations of, 56-57*t*, 59*f*
 see also Airborne dust
Suspended particulate matter emissions, 55*f*, 56*f*

Trace gases
 greenhouse effect of, 11*t*
 lifetime of, 14*t*
Traffic flow control, emissions and, 98*f*, 98-99, 99*f*
Traffic management. *See* Traffic flow control

Vehicle emissions. *See* In-use vehicle emissions
Vehicle user, road traffic emissions and, 96, 96*f*, 97*f*

Water vapor, global effects of, 13

About the Authors

Prof.Dipl.-Ing.Dr.Techn. Hans Peter Lenz was born in Bonn in 1934. Dr. Lenz studied mechanical engineering at the Technical University of Aachen and at the Federal Technical University in Zürich. He worked at various automotive companies for ten years. Since 1974, Dr. Lenz has been professor and director of the Institute for Internal Combustion Engines and Automotive Engineering at the Technical University of Vienna.

Dipl.-Ing.Dr.Techn. Christian Cozzarini was born in Vienna in 1969 and studied mechanical engineering at the Technical University of Vienna. From 1995 to 1999, Dr. Cozzarini was assistant at the Institute for Internal Combustion Engines and Automotive Engineering at the Technical University of Vienna. His work covered research on emissions and air quality, particularly emissions from road traffic.